THE CARRIER WAVE

New Information Technology and the geography of innovation, 1846–2003

PETER HALL
PASCHAL PRESTON

London
UNWIN HYMAN
Boston Sydney Wellington

Published by the Academic Division of
Unwin Hyman Ltd
15/17 Broadwick Street, London W1V 1FP, UK

Allen & Unwin Inc.,
8 Winchester Place, Winchester, Mass. 01890, USA

Allen & Unwin (Australia) Ltd,
8 Napier Street, North Sydney, NSW 2060, Australia

Allen & Unwin (New Zealand) Ltd in association with the
Port Nicholson Press Ltd,
60 Cambridge Terrace, Wellington, New Zealand

First published in 1988

British Library Cataloguing in Publication Data

Hall, Peter, *1932–*
The carrier wave: new information technology and the geography of innovation 1846–2003.
1. Telecommunication systems. Technological innovation, 1846–2003. Economic aspects
I. Title II. Preston, Paschal
384'.041

ISBN 0–04–445081–8

Library of Congress Cataloging in Publication Data

Hall, Peter, *1932–*
The carrier wave: new information technology and the geography of innovation 1846–2003.
1. Telecommunication systems. Technological innovation, 1846–2003. Economic aspects
I. Title II. Preston, Paschal
384'041

ISBN 0–04–445081–8

Typeset in 10 on 12 point Melior by Columns, Reading,
and printed in Great Britain by Biddles of Guildford

PREFACE

The research on which this book is based was supported by a grant from the Leverhulme Trust to the University of Reading. We should like to acknowledge with deep gratitude the support of the Trust and in particular its then Director, Professor Ronald Tress.

Many people in the University of Reading and in the Joint Centre for Land Development Studies gave unstintingly of their time and energy to help us in our work. In particular we want to acknowledge the late David Bishop, then co-Director of the Centre, whose help – on this as on other projects – went far beyond any ordinary contractual obligation, and his secretary Mary Tull; Gillian Bogue for her secretarial support; Sheila Dance and Chris Howitt for their cartographic services; and Rosa Husain who oversaw production of the draft and final manuscripts.

Nick Beavan was Research Officer in the early stages of the work, responsible for development of the statistical data bases. We want to acknowledge his invaluable help, without which this work would have been much the poorer. Equally, for help with preparation and processing of data, we wish to thank staff of the ESRC Data Archive staff at the University of Essex, and Alan Thornton of the Computer Centre in the University of Reading.

Finally, we owe a debt to the late Mr. L.A.W. Davis who checked the manuscript for its technological content; Professor Christopher Freeman, Dr. Ronald Abler and Dr. Roger Tooze for their helpful comments on the draft manuscript; and to Magda Hall for help in proof-reading. We must add the usual disclaimer: for errors and omissions, the authors take sole responsibility.

This book was in every sense a joint responsibility. Hall was Principal Investigator, Preston the Research Officer throughout the study. Preston was wholly responsible for the basic research and for production of the first draft of the manuscript, save for Chapters 8 and 13 and parts of 6 and 9, which were Hall's. Hall wrote successive drafts of the final manuscript in close consultation with Preston. The final version is the result of many hours of debate both as to theory and interpretation of the record. We are pleased, not to say profoundly relieved, that at the end we were able to agree to just about everything in the text now published.

Peter Hall
Paschal Preston
July 1987

CONTENTS

page
Preface iii
List of tables ix
List of figures xii

Part One *The Kondratieff waves*

1 Long waves, New IT and the geography of innovation
The ghost of Kondratieff 3
An alternative approach: New IT and the geography of innovation 5
The structure of the book 6

2 The long-wave debate
The contribution of Schumpeter 13
The revival of long-wave theory 16
Seven key questions 22
Our approach 25

3 New IT in a long wave perspective
IT: old, new and newest 29
Refining our hypotheses 30
The problems of data 33

Part Two *The mechanical age: New IT in the second Kondratieff, 1846–95*

4 The birth of New IT: telegraph and telephone
The first electrical innovations 38
The birth of the telegraph 39
The emerging cable industry 45
The invention of the telephone 47
The telephone in Europe 50
Conclusions 52

Part Three *The electrical age: New IT in the third Kondratieff, 1896–1947*

5 The electrical revolution
The basic innovations: batteries, generators, motors, transmission 58
The UK and the changing locus of innovation 65
Lighting, traction and other electrical applications 67
Creating a new industry 71

6 Information technology in the third Kondratieff: the mechanical byway and the electronic reunion
Electromechanical IT 75
Electronics: the scientific underpinnings 86
The working technologies 87
The birth of radio 89
The advent of broadcasting 94
Television 96
Radar 98

7 The origins of British decline
British electrical manufacturing in a world context 100
The growth of New IT industries 106
The causes of British failure 110
Small scale of enterprise 112
Failure in research and development 116
The rôle of finance 121
Conclusions 121

8 Electropolis: Berlin, Boston, New York and London, 1847–1952
Berlin: Elektropolis 124
Two revolutions: Berlin's electrical industry in the second and third Kondratieffs 126
The location of the industry within Berlin 129
The fall of Elektropolis 136
Boston to New York: the emerging electrical Megalopolis 137
The comparative case of London 145

Part Four *The electronic age: New IT in the fourth Kondratieff, 1948–2003*

9 The electronic revolution
The transistor: core technology 151
The birth of the computer 154
Advances in telecommunications 160
The consumer electronics sector 162

Office technology in the fourth Kondratieff 163
Conclusions 164

10 New IT on the world stage: the fourth Kondratieff upswing
The office equipment industry 167
Electronics capital goods 173
Telecommunications equipment 179
Electronic components and semiconductors 181
Conclusion 185

11 Information technology in the world economy: the fourth Kondratieff downswing
The geography of information technology 189
Measuring employment change, 1970–81 189
Changes in individual New IT categories 195
Employment and output in the USA, UK and Japan, 1958–82 199
International trade in New IT 203
Summing up 212

12 The British anomaly
Employment and output in the third and fourth Kondratieffs 216
The anatomy of decline 222
Towards explanation: the rate of innovation 232
The R&D factor 234
The military bias 236

13 The flight from Electropolis: innovation on the move
The fall of Elektropolis 243
London's Western Corridor 244
The American case: the Boston–New York Corridor 253
Three New IT agglomerations: Route 128, Silicon Valley and Orange County 255
Conclusion: the geography of innovation 260

Part Five *The information age: Convergent IT in the fifth Kondratieff, 2004–*

14 Explanations and speculations
The rôle of basic innovation 265
Five technology systems, and their diffusion 268
The rôle of infrastructure 273
Economies of scale in New IT 274
The geography of innovation 275
Innovation and industrial organization 277
The new rôle of the state 278

Defence-led innovation in IT 279
The significance of societal innovation 280
Towards the fifth Kondratieff 284

References 289

Index 302

List of tables

2.1 The four Kondratieff waves: different chronologies 19
2.2 The four Kondratieff waves 21
2.3 Four categories of innovation 23
3.1 Definitions of New IT, by SIC categories 31
4.1 UK Post Office telegraph revenues and messages handled, 1856–1 and 1911–2 44
7.1 Employment in electrical apparatus and supply occupations, England and Wales, 1881–1911 101
7.2 Net output and employment in UK electrical engineering industries, 1907–35 101
7.3 Value of electrical manufacturing production, main countries, 1907, 1913, 1925 102
7.4 Exports of electrical machinery and apparatus, five major countries, 1913–28 102
7.5 Germany: employment in electrical and all industry, 1875–1965 103
7.6 UK electrical undertakings, 1900–36 105
7.7 UK: major radio valve producers, 1919–39 108
7.8 Radio equipment: exports, leading countries, 1923–7 109
8.1 Leading centres of the German electrical industry, 1925–39 125
8.2 Employment in S&H and AEG, 1880–95 125
8.3 Location of the Berlin electrical industry, 1895–1939 132
8.4 Spinoffs from Telefunken, 1903–28 134
9.1 Leading US semiconductor firms, 1950–79 155
10.1 Shares of main producing countries in output of ‘other’ office machinery, 1954–66 168
10.2 Exports and imports of office machinery, UK, 1954 and 1968 168
10.3 Estimated sources of office machinery supply in the UK, 1964–6 170
10.4 Production of electronic capital goods, national shares, 1964 178
10.5 Telecommunications equipment: estimated production and exports, major countries 1958–64 180

10.6 Electronic component production, major countries, 1965 185
11.1 Definition of New IT industries on the basis of the ISIC 190
11.2 New IT: top 12 nations ranked by absolute employment growth, 1970–81 193
11.3 New IT: top 12 nations ranked by percentage employment growth, 1970–81 194
11.4 New IT: top 12 nations ranked by location quotient, 1970, 1976, 1981 195
11.5 New IT trade product groups: definition 202
11.6 Office machinery etc. (SITC 75): leading export nations 205
11.7 New IT exports: composition by sector, 1982 211
11.8 New IT: net trade balances, 1982 211
12.1 New IT industries, UK: average annual growth rates, 1948–81 218
12.2 New and Old IT industries and services: employment, 1981–4 220
12.3 New IT products: UK trade balances, 1982–4 220
12.4 Electronic capital goods, estimated deliveries, UK, 1958–65 225
12.5 Major UK producers of valves and cathode-ray tubes, 1954 229
12.6 Semiconductor firms in Great Britain and years they were active in the industry, 1954–68 230
12.7 UK semiconductor industry: leading firms, 1962 and 1973 231
12.8 US patents granted to inventors from selected countries, 1966–82 232
12.9 US patents granted in IT-related fields, 1963–81 233
12.10 Major country shares of selected foreign patenting in USA, 1981 233
12.11 Estimated civilian R&D ratios for selected countries, 1961–78 234
12.12 National expenditures for R&D, 1969 and 1979 234
12.13 Selected R&D indicators, major OECD countries, c.1983 235
12.14 Changes in UK intramural industrial R&D expenditure, 1975–83 236
12.15 Estimated civilian and military R&D, major OECD countries, 1983 237
12.16 Shares of UK state-funded R&D by major departmental grouping 237
12.17 State R&D most related to specific industrial sectors, 1983–4 238

12.18 Industrially performed R&D, UK, by major product groups 239
13.1 Percentage regional distribution of total employment in New IT producing industries, GB, 1881–1981 245
13.2 Employment change in New IT industries, GB, 1971–81 246
13.3 GB regional distribution of employees in employment in New IT industries (1981) ranked according to location quotients 247
13.4 GB regional distribution of total employees in employment in New IT industries (1981) by MLH ranked by total employment size 248
13.5 USA: New IT jobs, 1977 and 1981 253
13.6 USA: New IT employment, by leading states, c.1977 253
14.1 Major New IT innovations: dates and countries of origin 267
14.2 Electronics innovation chains 268

List of figures

8.1 Greater Berlin: electrical industry, key locations, 1842–1921 130
8.2 New York City region: electrical industry, key locations, 1869–1942 139
11.1 Employment change, New IT industries, world estimate, 1970–81 191
11.2 Employment, New IT industries, selected countries, 1970–81 192
11.3 Employment, office machinery, selected countries, 1970–81 196
11.4 Employment, telecommunications, components, radio and TV equipment, selected countries, 1970–81 197
11.5 Employment, professional and scientific instruments, etc., selected countries, 1970–81 198
11.6 Employment, New IT industries, USA, UK, Japan, 1958–82 200
11.7 Net output, New IT industries, USA, UK, Japan, 1958–82 201
11.8 Exports, New IT, world estimate, 1971–82 203
11.9 Exports, New IT, per cent of estimated world total, 1971–82 204
11.10 Exports, computers and other office machinery, per cent of estimated world total, 1971–82 206
11.11 Exports, telecommunications and sound recording apparatus, per cent of estimated world total, 1971–82 207
11.12 Exports, electronic components, semiconductors, etc., per cent of estimated world total, 1971–82 208
11.13 Exports, photographic, etc., apparatus, per cent of estimated world total, 1971–82 209
11.14 Exports, professional and scientific instruments, etc., per cent of estimated world total, 1971–82 210
12.1 Employment, New IT, UK, 1907–81 216
12.2 Employment, constituent New IT industries, UK, 1907–79 217
12.3 Employment and net output, all manufacturing and New IT industries, UK, 1948–81, 1935–81 219

12.4 Ratios of imports to home demand, New IT industries, UK, 1963–82 221
12.5 Sectoral composition of output, electronics industries, UK, 1954–64 224
12.6 Computer manufacturers, UK, 1950–70 226
12.7 Electronics exports and imports, UK, 1958–65 229
13.1 New IT industries, employment location, UK, 1981 249
13.2 New IT industries, employment location gains and losses, UK, 1971–81 250
14.1 Schematic diagram of radical New IT innovations and the Kondratieff long waves 269

Carrier . . . n. Telecomm. An electromagnetic wave or an alternating current or voltage that is modulated by a wave, etc., of lower frequency representing the signal to be transmitted . . . 1931 *B.B.C. Year–Book* 437/2 *Carrier Wave*, the high–frequency oscillations emitted by a wireless telephone transmitter. These are modified during telephony.

Burchfield, R.W. (ed.) 1972. *A Supplement to the Oxford English Dictionary*, Vol. I. Oxford: Oxford University Press.

PART ONE

The Kondratieff waves

1

Long waves, New IT and the geography of innovation

Long waves of economic development, the starting point and main organizing concept of this book, have long been the subject of impassioned intellectual controversy. Yet – whatever their cause, whether or not they even exist – on one point there can be no debate: there is a definite long wave in academic interest in them. And it runs a countercyclical course: booming in the recessive 1920s and depressive 1930s, moribund in the expansionary 1950s and 1960s, recovering in the recessionary 1970s, flourishing again in the depression-ridden world of the 1980s (Mandel 1980; van Duijn 1983). The reason, evidently, is that the obsessions of social scientists mirror the state of the world outside the ivory tower: in periods of economic upswing, economists study the conditions of economic development and geographers produce models of growth; in periods of recession, economists rediscover the business cycle while geographers analyse the anatomy of job loss. And both turn to the study of innovative, high-technology industry, which has become '*the* economic development strategy of the 1980s . . . the only economic development game in town' (Saxenian 1985c, 121).

The ghost of Kondratieff

So, neglected and almost forgotten for a third of a century, the name of Nikolai Kondratieff has returned to haunt the halls of academe. It is easy to see why. Kondratieff, an early Soviet economist whose classic papers were published in the mid-1920s, certainly was not the first to discover long waves; but he did use them to provide both an explanation of why capitalism should experience major economic crises at approximately half-century intervals, and an analysis of the way in which it would – just as predictably – escape from them (Kondratieff 1935). Perhaps, though the ostensible excuse was that he was trying to organize an opposition party, that was the real reason why Stalin had him arrested and sent to Siberia; there, according to Solzhenitsyn's *Gulag Archipelago*, he died insane (van Duijn 1983).

Long waves, according to Kondratieff, were accompanied by changes in technology, but were not triggered by them; for new techniques could be exploited only when the economic conditions were right. But a decade after Kondratieff wrote, Joseph Schumpeter came to apply his theory in detail, and stood it on its head (Schumpeter 1939). He argued that the start of each long wave was indeed caused by a new bunch of innovations, which created whole new industries: coal, iron and steam in the first Kondratieff (1785–1842), steel and railways and steamships and machine tools in the second (1843–97), cars and electrical goods and chemicals in the third (1898–). So, in the subtitle of the book by Gerhard Mensch (1979), Innovations Overcome the Depression. This, of course, represented a direct challenge to the conventional Keynesian, and behind that the entire neoclassical, economic tradition, which had almost totally ignored the rôle of technological change; and the challenge came just when some economists, and many policy makers, were becoming disenchanted with Keynesian analyses and remedies. Small wonder that in the 1980s interest in Kondratieff and Schumpeter should revive; and nowhere more than in the UK, long the sick man of the capitalist world.

The resulting flood of literature has thrown much empirical light on the waves, but has generated no agreement as to their dating or their causes. Among the researchers, there is now at least general consensus as to the existence of a long wave of some 50–60 years' duration. There is however much difference in detail on their precise timing, with a special degree of disagreement about the current, fourth, Kondratieff. There is also some agreement that technological innovation plays a rôle. But there is very fierce debate as to the causal relationships. Does innovation really peak at approximately half-century intervals? Is it, so to speak, autonomous, surging at regular intervals according to laws of its own? Or does it respond to crises of capital accumulation? What is the rôle of organizational and, indeed broader social innovations? Do innovation waves depend on strictly contingent and random events, like wars? But do not wars also follow a long-wave pattern, as Kondratieff asserted? All these questions are still exciting sharp controversy and a flood of literature.

An alternative approach: New IT and the geography of innovation

Much of this debate has taken place at a very high level of historical generality. Other, more detailed, work has had a restricted time-span. In this book we try to take a different approach: we look at one complex of technologies and industries over the span of three Kondratieff cycles. We have chosen the group of industries that during this long period – not least, during the recent fourth Kondratieff downswing – have increasingly been at the leading edge of technological innovation, and may fairly be called the key industries of the current fourth Kondratieff downswing: the so-called New Information Technology (hence, New IT) industries, embracing the technologies – mechanical, electrical, electromechanical, electronic – that record, transmit, process and distribute information.

New IT, as will emerge, was born during the 19th century. For centuries before that there had been information technologies: writing, printing, simple forms of telegraphy. But the invention of the electrical telegraph in the 1830s, and that of the telephone and the typewriter and the phonograph in the period 1875–1890, marked a distinct new era of technological development: hence the term New IT, NIT for short. More recently, in the 1970s and early 1980s, we have witnessed another burst of innovation which is increasingly producing a convergence of formerly separate New Information Technologies, particularly the computer and the telephone. Some literature refers to this latest development as 'New IT'. Since for us the term refers to the cluster of information technologies that began in the 1870s and continued through the third and fourth Kondratieffs, we shall call the newer development 'Convergent IT', CIT for short.

Our aim in studying the genesis and development of the New IT industries has been twofold. First, we hope that thereby we shall throw greater light on the precise nature of the innovative process and on its historical contribution to the creation of new industrial traditions, particularly the impact of major new technological systems on economic and social development. But that is ancillary to our second and main purpose: to understand the changing geography of innovation. Evidently, over the course of capitalist history the fortunes of the main industrial nations have waxed and waned. In the first Kondratieff, Britain was pre-eminently the major, almost the sole, industrial nation. In the second, it was increasingly challenged by Germany and the USA. In the third, these two challengers took the lead; by its end, the United States was indisputably world industrial leader. During the fourth, it retained its leadership but was increasingly challenged by a relative newcomer, Japan. Precisely the same sequence applies for the New IT industries, which first appeared in the second Kondratieff.

And, just as on the international stage, so within each of the leading national economies: the locus of the leading-edge innovative industries has switched from region to region, from city to city. From the birth of New IT in the second Kondratieff and on through the third, Berlin and the Boston–New York Corridor were the main global centres of innovation; during the fourth, they were supplemented or supplanted by new urban centres such as Southern California, Silicon Valley, the Western Crescent around London, the Stuttgart–Munich Corridor, and the Tokaido Megalopolis. Our central question is why these shifts should have occurred. Do they truly reflect changes in innovatory energy? And if so, what forces could be responsible?

The structure of the book

The book has a simple structure, evident from the table of contents. It is divided into five parts.

Part One is introductory to the main subject matter. Chapter 2 aims to provide a succinct overview of the burgeoning literature on long waves and the main lines of the long-waves debate. It further develops the point that right from the start there were important differences – not least between Kondratieff and his leading exponent, Schumpeter – on the causal mechanisms. It documents the criticisms by Kuznets and others, during the 1940s and 1950s, of the long-wave concept, including the suggestion of a shorter (20–25-year) construction cycle: the Kuznets cycle. It summarizes the work of Mensch, which essentially relaunched the whole debate during the 1970s, and examines the detailed criticisms of his thesis by the Science Policy Research Group and others. Finally it tries to sum up the state of the debate at the time of finishing this book, Autumn 1986.

Chapter 3 turns to the New Information Technology industries themselves. It seeks to isolate their distinguishing characteristics, especially *vis-à-vis* the older information technologies which they supplemented and partly replaced. It discusses some of the problems of defining and measuring them, especially across international boundaries and across decades. It concludes with a summary of the main technologies and their evolution over the long period from 1842 to the present: the period of the so-called second, third and fourth Kondratieffs.

Thereafter, the book is organized chronologically in three parts, corresponding to these three long waves. The treatment becomes increasingly dense and detailed as we move from one part to the next. That is first because the necessary information, especially the necessary statistics, becomes increasingly plentiful as we move towards the present day; secondly because the NIT industries play an increasingly important role, marginal in the second Kondratieff, significant but indirect in the third, increasingly dominant (in the form of CIT) during the fourth Kondratieff downswing. In the fourth Kondratieff, indeed, they begin to play a role like steel and ships in the second, transformers and cars and pharmaceuticals in the third: by the downswing of that long wave, in a Schumpeterian sense, they become the true technological carriers of the wave.

The main focus in these chapters is on a comparison between the major nations producing New IT: the UK, Germany, the USA, Japan. And in particular – given that the research was conducted in the UK, against a background of intense interest in the future of high-technology industry there – it puts the UK, and its relative competitive failure, front-of-stage. But an important subsidiary theme, closely linked to the international comparison, is the regional and urban locus of innovation.

Chapter 4, forming a single-chapter part on the second Kondratieff, outlines the origins of New IT at the start of this long wave, in the form of the electric telegraph. Discovered almost simultaneously in Great Britain and the USA in the 1830s, it was the first electrical innovation. It soon became the first agent of instant communication between countries and continents, engendering a new complex of industries which made the pieces of apparatus and above all the cables which connected them. The UK and the USA were the logical seats of these industries, though Germany soon began to challenge them. It also triggered a series of further innovations to improve the system and develop specialized applications, above all in the developing financial markets of the world, associated with the name of Thomas Edison. These, plus the discovery of the telephone by Alexander Graham Bell, a Scots emigrant to the USA – gave that country a pronounced innovative lead towards the end of the second Kondratieff – a lead it never afterwards lost.

Chapter 5, the first of four on the third Kondratieff, represents a massive but necessary diversion from the main story. It recounts the birth of the modern electrical industries, arising from an extraordinary cluster of innovations in the generation and transmission of heavy power during the 1870s and early 1880s. These came especially from Germany and above all – thanks again especially to Edison – from the USA. In turn they almost instantly created new industrial giants, organized from the start on a huge scale and depending on massive research and development, in these three countries: carriers, together with cars and chemicals, of the third Kondratieff which is dated from the second half of the 1890s.

The reason why this is a necessary diversion is explained in Chapter 6: the new electrical technologies, though they might seem natural partners of NIT, were not effectively married to it for at least another three decades. Instead, a whole series of further innovations of the 1870s and 1880s – many associated with Americans like Edison, others coming from Germany – brought a whole range of New IT industries depending on mechanical, not electric, power: the typewriter, the gramophone, the dictation machine, the rotary stencil duplicator. Only after the development of effective valves in the first decade of the 20th century, making possible wireless telegraphy and ushering in the electronic age, was electric technology again applied to the task of transmitting information. And, down to the 1930s, the radio industry, a belated product of the third Kondratieff, was almost the sole industrial outcome. Consequently, though this Kondratieff could fairly be called the electrical age, in much of it New IT went down a mechanical byway.

In Chapter 7 the study turns to consider the geographical implications, first on the international stage. It poses the question: Why, during the third Kondratieff, did the UK suffer such a massive competitive setback? For, right from the start of this long wave, the UK's major emerging competitors – Germany and the USA – wrenched world leadership from her. Partly that was because many of the basic innovations were made there; partly it had to do with their subsequent exploitation. In both those countries the new electrical industries were organized from the start in giant oligopolistic companies, based partly on links with major financial houses. These provided the large scale investment necessary for the more rapid diffusion of the new technology and joint development of the electrical supply and equipment industries, and which soon reached agreements to regulate and to share all or part of the market. This was particularly evident after the turn of the century in radio, where at the beginning the UK took a substantial lead through the Marconi company, but where first Germany and then the USA reacted by trying to create virtual monopolies. This marks the fact that, increasingly, the state was beginning to intervene in the development of these industries: first in Germany, later in the USA, it encouraged mergers, regulated competition, promoted education and research, and finally, through its defence ministries, began to become a major customer for high-technology products. In all these respects the UK, still wedded to a 19th-century *laissez-faire* concept of the state's rôle but perversely wedded to archaic regulation and older forms of industrial structure, progressively fell behind: such is the thesis of the chapter.

Chapter 8 develops this theme on the regional and local scale. In the third Kondratieff the great national metropolitan regions (Berlin, New York–Boston, London) were the major seats of global innovation in both electrical and New IT technologies: the Silicon Valleys of their day. The reasons were complex: in Germany above all, the demands of government and especially of the war machine; in New York especially, the demands of finance capitalism; in both Berlin and Boston, the concentration of high-quality technological education. As R&D increasingly became concentrated in large, well organized industrial laboratories, financed out of profits, these economies of urban agglomeration became less pressing; such laboratories, as the economists say, internalized their externalities. Some of them were thus able to move out of the metropolitan milieu, attaching themselves to production centres like Schenectady or Pittsburgh; but others, on the model of Edison's pioneering venture of the 1870s at Menlo Park in New Jersey, stayed within the general sphere of the great metropolis, moving at most to its rural fringes. And again, London, though it remained Britain's indisputable centre of high-technology research and production, limped behind its major metropolitan competitors.

Chapter 9 brings us to the fourth Kondratieff, the origin of which is generally dated as the late 1940s. Paralleling Chapter 5, it outlines the sequence of critical innovations that ushered in the electronic age. The central technology is of course the computer in all its manifestations. But the computer is an assemblage of electronic components; crucial to its effective and efficient operation, and therefore the key technology on which all else rests, is the semiconductor, first developed in the form of the transistor at Bell Laboratories in December 1947. As it came into mass production it was initially applied to existing electronic products – radio, above all television – which themselves were carriers – minor ones, admittedly – of the subsequent economic upswing. But parallel to that, the transistor and its successor technologies – the integrated circuit, the microprocessor – were embodied into exotic aerospace products, produced to the demands of government agencies in response to the missile and space races that began in the mid-1950s and continued, sometimes waxing, sometimes waning, through the following decades. Above all in the USA they thus created a vast complex of new industries, sometimes producing large civilian technological spinoffs in their wake.

Chapters 10 and 11, accordingly, take up the story of these industrial repercussions. Because by now the story is denser in detail, it comes in two halves: one dealing with the growth of the new industries in the expansionary phase of the 1950s and 1960s, the other with the challenges they faced in the downswing of the 1970s and early 1980s. The main question, again, concerns the changing fortunes of the main industrial economies. The USA, as might be expected, starts from a pre-eminent position and, bolstered both by a huge lead in innovation and a burgeoning military demand, reinforces it. Europe, which at the end of World War II seems to be the major challenger, progressively falls away. In the case of Germany this can be explained partly by the catastrophe of 1945, from which the country's long tradition of New IT research and development never quite recovered. In the case of the UK, which had actually made up much of the ground lost in the earlier decades of the previous Kondratieff, the relapse remains a mystery awaiting further analysis. What is evident is that Europe's, and above all the UK's, loss is eastern Asia's gain: Japan in the 1960s, the Asian newly industrializing countries in the 1970s and 1980s, now emerge as the USA's major challengers.

Since the British failure remains a puzzle, Chapter 12 takes it up. It first documents the continued loss of markets in one after another of the major New IT sectors, confirming that far from compensating for the UK's decline in the more traditional industries, New IT has perversely shared their fate. Then it goes into causes, reaching conclusions that in many respects contradict the comfortable conventional wisdom; in particular, it appears that by the fourth Kondratieff the UK was no longer a significant industrial innovator. So the chapter goes on to ask why. It concludes by posing the question: Are there any lessons from the British experience for other mature industrial nations, above all for the USA, whose recent evolution shows some disturbing similarities?

Chapter 13, concluding this section, parallels Chapter 8: it looks at the changing regional and urban locations of the New IT industries during the fourth Kondratieff. This long wave was marked by major geographical shifts: the abrupt, even catastrophic, decline of Berlin, once electrical capital of the world; the movement of the New IT industries from London to its neighbouring Western Crescent; the continuing strength of the Boston–New York axis but also the rise of whole new high-technology complexes in Southern California and Silicon Valley. The chapter reviews the alternative explanations, advanced in the literature, to reach a view on the forces that help keep such industries in their traditional locations, or alternatively help to trigger the development of whole new industrial regions.

Finally, Chapter 14 sums up. It tries to review the evidence for the links between innovation and industrial growth. Also it asks whether there are useful or consistent generalizations that can be made about the changing locus of innovation. In particular, it concentrates on some major themes that emerge both in this study and in the general literature: the rôle of industrial organization and industrial scale, the precise nature and significance of state policies, the part played by demand and thus by the evolution of the general economy outside the narrowly defined high-technology sectors. Thus it returns to a central question posed in Chapter 2: How far do bunches of industrial innovations create not merely new industries, not merely even complexes of related industries, but also a whole new technological–industrial paradigm that transforms the entire nature of economic and social processes and relationships? This brings it to a final question: What are the likely carriers of the coming fifth Kondratieff, what will be their impact on the structure of economic life, and what does all this say for the competitive prospects of the British and American economies?

2
The long-wave debate

Strictly, the Kondratieff waves are a misnomer. It was Schumpeter, in his two-volume *Business cycles* of 1939, who thus christened them (Schumpeter 1928, 1939). But Schumpeter, who spent a lifetime accumulating the materials for his monumental, posthumously-published *History of economic analysis*, should have known better than anyone that this represented an historical injustice: Kondratieff, who certainly thought that he had discovered them as early as 1919, recognized in a footnote that he had first learned in 1926 of the earlier work of the Dutch economists de Wolff and van Geldern (Kondratieff 1926, 1935, 1984; van Duijn 1983). Yet Schumpeter mentions van Geldern not at all, and de Wolff only in another context (Schumpeter 1954). So the name has stuck; and there is at least this justice, that Kondratieff's monumental labours of the 1920s – the great bulk still solely in the original Russian – represent the definitive early scholarship on the subject of the long waves.

It is however empirical rather than theoretical, relying on time-series analysis to relate price movements to indices of industrial activity; indeed, Kondratieff was heavily criticized by fellow Soviet economists of the 1920s for his lack of explanation (van Duijn 1983). In so far as there is one, it is not the one that Schumpeter later made the main plank of his own adaptation of Kondratieff: the bunching or swarming of innovations. Rather, Kondratieff relied on variations in the renewal of capital goods, related to the price of loan capital, an explanation that suffers from a fundamental defect, that it cannot readily explain why investment should start up again at the bottom of the cycle (van Duijn 1983).

There is, however, a very good reason why Kondratieff's theory should have run foul of official ideology in the Soviet Union; whatever his explanation, however impeccable the derivation from other Marxist theoreticians, even from Marx himself, the fact was that the theory suggested an endogenous and automatic mechanism, whereby the capitalist economy would regenerate from each successive crisis, rather than the final cataclysm for which the Soviet leaders of the 1920s were waiting (van Duijn 1983). Indeed, Trotsky – the ultimate deviationist – had shared Kondratieff's notions of a wave-like movement of the capitalist economy (van Duijn 1983). Nevertheless, the remarkable fact is that Kondratieff, having identified innovation swarms as one feature of the economic upswing, failed to see that this would provide the explanatory mechanism he needed. He had noticed, in his classic 1926 paper, that

> Changes in the technique of production presume (1) that the relevant scientific–technical discoveries and inventions have been made, and (2) that it is *economically* possible to use them . . . Scientific–technical inventions in themselves . . . are insufficient to bring about a real change in the technique of production. They can remain ineffective so long as economic conditions favorable to their application are absent . . . the development of technique itself is part of the rhythm of the long waves. (Kondratieff 1926, 593–4; 1935, 112)

This provides the other half of the theory he needs: at a certain point in the capital-replacement cycle a bunch of inventions is available to be turned into commercially marketable innovations, thus triggering a wave of investment. Though he is not explicit on the point it seems that this process is endogenous, requiring no outside intervention. In this regard, Kondratieff in the 1920s had gone beyond the macro-economic literature of the 1950s, though he did not fully appreciate it (van Duijn 1983).

The contribution of Schumpeter

Schumpeter of course *did*. That is why he inverted Kondratieff: the inventions come first, but they are not yet innovations; the entrepreneur, 'the carrier-out of new combinations', has to perform that job of conversion; and the innovations trigger the upswing. He thus overcame the deficiency he found in neoclassical economists, including even Keynes, whom he regarded as perhaps the greatest neoclassicist of all: their obsession with short-term equilibrium and their total neglect of the longer-term dynamics of capitalist growth (Schumpeter 1952, 1954). He also wrote the uncompleted chapters of the economist whom he regarded as an even greater master, the one who had grasped the dynamics of capitalism, Marx: Schumpeter supplied the missing Marxian connections between the theory of the capitalist cycle and the theory of capitalist crisis, and made the distinction he believed Marx had failed to make, between the capitalist and the entrepreneur (Schumpeter 1942, 1954).

Schumpeter's resulting theory ran like this. Neoclassical economics, including the Keynesian variant, could explain only the self-equilibriating changes that occurred in the 'very strictly short run': a few years at most, corresponding perhaps to the shortest of the cycles that Schumpeter recognized, the 40-month or Kitchen cycle (Schumpeter 1952, 1954); thus, it restricted itself 'to the factors that *would* govern the greater or smaller utilization of an industrial apparatus *if* the latter remains unchanged' (Schumpeter 1952, 287). But he went on to argue:

> Those who look for the essence of capitalism in the phenomena that attend the incessant recreation of this apparatus and the incessant revolution that goes on within it must therefore be excused if they hold that Keynes's theory abstracts from the essence of the capitalist process. (Schumpeter 1954, 1175)

Here, Schumpeter followed the creative insight that he had first developed in his *Theory of economic development*, written when he was only 24. The system described by the neoclassicists was a 'circular flow' that was essentially static: it would tend to equilibrium as long as circumstances did not change (Schumpeter 1961), 'running on in channels the same year after year' (ibid., 61). The real question was what caused change. Schumpeter argued that change came from within the system (Schumpeter 1961), via what he called 'the carrying out of new combinations': a new good, a new method of production, a new market, a new source of supply, a new industrial organization all represented such 'new combinations' (ibid.). It was these that constituted what he later called the process of 'creative destruction', the essential feature of capitalist development (Schumpeter 1942). The 'carrier-out of new operations' was the entrepreneur (Schumpeter 1961); he was, however, not the risk-taker, who was the capitalist (ibid.). Already, at this point, Schumpeter had applied his theory to explain the cyclical character of the capitalist development process:

> The boom ends and the depression begins after the passage of the time which must elapse before the products of the new enterprises can appear on the market. And a new boom succeeds

> the depression when the process of resorption of the innovations is ended. (Ibid., 213)

Further, he already suggested:

> The new combinations are not, as one would expect according to general principles of probability, evenly distributed through time . . . but appear, if at all, discontinuously in groups or swarms. (Ibid., 223)

Such swarming, he argued, was due to the fact that the very appearance of entrepreneurs encouraged others to imitate them, in ever increasing numbers; thus, there would be an increase in the rate of investment and the growth of capital-goods industries (ibid., 230).

What is interesting for the historian of economic theory, is that at this early date (about 1907) Schumpeter was thinking only of the relatively short business cycles, of about nine years' duration, first identified by the French economist Clément Juglar (Schumpeter 1954). Juglar's analysis had led him to his famous aphorism that 'the only cause of depression is prosperity' (ibid., 1124); the problem then was to explain the prosperity, to which he had no answer (Schumpeter 1954). Schumpeter claimed to provide it. But it was of course only 20 years later that he discovered the Kondratieff waves and embodied them in the more complete theory that he published in 1939.

This argued that there were at least three business cycles: the Kitchen of some 40 months, the Juglar of just over 9 years and the Kondratieff of some 57 years' duration. They nested one within the other, 'six Juglars to a Kondratieff and three Kitchens to a Juglar – not as an average but in every individual case' (Schumpeter 1939, 173–4). Schumpeter's own statistical analysis led him to conclude that each cycle followed 'three sine curves with periods respectively of 684, 114 and 38 months (or 57, 9½ and 3⅙ years) and with amplitudes roughly proportional to the periods, i.e. in the relation of 18, 3, and 1' (ibid., 1051). The dates he gives for the start of the first three Kondratieffs – 1787, 1842/3 and 1898 – actually represent lengths of 56, 54 and 56 years, respectively, though he stresses that these dates are 'tentative' and 'approximate' (Schumpeter 1939).

Each of these – here Schumpeter is specific – represented a separate industrial revolution resulting from the arrival of a new group of technologies:

> Historically, the first Kondratieff covered by our material means the industrial revolution, including the protracted process of its

absorption. We date it from the eighties of the eighteenth century to 1842. The second stretches over what has been called the age of steam and steel. It runs its course between 1842 and 1897. And the third, the Kondratieff of electricity, chemistry and motors, we date from 1898 on. (Ibid., 170).

This is highly significant. The 'new combinations' could take forms other than technological: innovation, for Schumpeter, was 'the setting up of a new production function. This covers the case of a new commodity, as well as those of a new form of organization such as a merger, of the opening up of new markets, and so on' (ibid., 87). But it is clear that in his view the Kondratieff triggers, the 'innovations of relatively long span' (ibid., 167), were new technologies that created new industries. In the long subsequent treatment, where he analyses the history of capitalist evolution in meticulous detail from the industrial revolution to his own time, this is the major structural theme.

It was of course criticized. Almost immediately on appearance *Business cycles* met with a barrage of criticism, both on statistical grounds and on the basis that the causal mechanism was insufficiently substantiated. One of the weightiest critiques came from Simon Kuznets, who had already spent a decade on the analysis of economic time series (Kuznets 1940). This and subsequent work led Kuznets to specify that the longest recognizable cycle was one of 20–25 years' duration, associated apparently with variations in activity in the construction industry. Meanwhile, in the prosperous decades of the 1950s and 1960s, interest in long waves waned; Schumpeter himself was reduced to the status of a 'footnote economist' (van Duijn 1983). The Keynesian version of neoclassicism reigned supreme; and in it, technological change was treated as a residual or relegated to a black box; in so far as it was mentioned, it was seen as incremental and regular (Rosenberg 1982, van Duijn 1983, Clark 1985).

The revival of long wave theory

The revival in long wave studies in the late 1970s, as suggested at the start of this book, partly reflected a natural reaction to the events of the day. Keynesian macro-economic management, thought more or less infallible for three decades, was now failing to deliver the expected results; throughout the capitalist economy a sharp reduction in the rate of economic growth was observable; in many countries it was bewilderingly associated with an explosion in money wages and prices, producing a new phenomenon of stagflation. A whole generation of economists, dissatisfied with what they saw as the narrow assumptions and artificial simplifications of Keynesian neoclassicism, began to look outside its confines. Many turned to Marxism for a more complete account of capitalist evolution, which related economic to social and political change in a way that the neoclassical framework never could: political economy, buried since the time of Jevons and Marshall, now reappeared. It was only natural that Marxian economists should rediscover Kondratieff and some earlier pioneers in the tradition. But others, grappling with the deficiencies of neoclassicism, also argued that technology must be brought out of the 'curious sort of underworld' where it had dwelt for too long (Rosenberg 1976); and that would demand going outside the formal limits of economics.

This revival was strongly centred in the Low Countries and in Germany, where interest in both Schumpeter and in long-wave studies had continued to flourish even when it was moribund in the Anglo-American world. A group in the Netherlands pursued research on long waves; one of them, Jacob van Duijn, later produced one of the standard English-language works on the subject. The Belgian Marxist, Ernest Mandel, made long waves a central feature of his book *Late capitalism*, first published in German (and based on seminars taught in Berlin) in 1972. Finally in 1974 a German scholar, Gerhard Mensch, published his study on innovation, which in many ways sparked the revival of interest in the subject.

Published in English translation in 1979, *Stalemate in technology: innovations overcome the depression* follows a Schumpeterian framework: it systematically analyses the pattern of invention and innovation from the 18th century to the present day, to conclude that indeed innovation has occurred in a strongly bunched pattern. It peaked sharply in the so-called radical years of 1764, 1825, 1886 and 1935 – between 11 and 17 years before the start of a Kondratieff cycle. Mensch further shows that the basic inventions, from which these innovations stem, do not show the same marked tendency to peak; he concludes from this that there is a so-called wagon train event, whereby inventions have to wait for suitable economic conditions before entrepreneurs arrive to convert them into marketable products.

The sensational impact of Mensch's book arose mainly from the fact that he used his analysis not merely to point to a pattern in past history but to forecast its course in the future. He confidently predicted that the next innovation wave would begin in 1984 and would peak in 1989, suggesting that the fifth Kondratieff would begin sometime around the year 2010. This, coming at a time when the capitalist world was just descending into the worst recession for 50 years, could hardly fail to make an intellectual impact. Much of the subsequent explosion of long-wave studies in the Anglo-American world has centred on a reaction to his work.

Just as with the reaction to Schumpeter in the 1930s, much of it has predictably taken the form of technical criticism. The weightiest has come from Christopher Freeman and his colleagues at the Science Policy Research Unit (SPRU) at the University of Sussex, who pioneered the study of long waves in the UK (Freeman *et al.* 1982, Freeman 1984, 1985, Rothwell & Zegveld 1981). They argue that his data are incomplete and biased, both as to selection and dating, and that analysis of a fuller list of innovations, drawn from the same source, shows no such innovatory swarming (Freeman *et al.* 1982). They tentatively accept the existence of long waves, but incline to share Mandel's interpretation: that there is no single triggering mechanism, certainly no automatic one, endogenous to the economic system. Rather, individual long waves appear to be initiated by unique exogenous events such as gold discoveries, wars or major social or institutional changes. More recently, a weighty technical criticism has come from Solomou, who concludes that Mensch's results are faulty on at least two counts: an unrepresentative choice of innovations, and an inappropriate statistical test. He thus supports the SPRU group's conclusion that 'the flow of innovation is not responsive to long-wave phases' (Solomou 1986, 111). But other work, including the very thorough analysis of an extended list of innovations by van Duijn, and Kleinknecht's testing of equally comprehensive data from Mahdavi, both conclude that depression-triggered innovation bunching has occurred (van Duijn 1983, Kleinknecht 1984). SPRU researchers have sophisticated the concept of the long wave by suggesting that a particular innovation, the automobile for instance, may act as a carrier of more than one wave, diffusing from one country or continent to another as it does so (Freeman *et al.* 1982).

At the same time, other economists have attacked the theoretical foundations of the long wave. Rosenberg and Frischtak have argued that the theory must be shown to meet certain rigorous, interdependent criteria. It must show that innovation is the causal agent, and explain its timing; it must then demonstrate the resultant backward and forward linkages through the economic system, and finally show why the process should recur. Like Freeman and his colleagues, Rosenberg and Frischtak incline to the view that long wave upswings may result in part from exogenous forces, in part from the revival of older technologies rather than the impact of new ones; thus the recovery of the late 1930s was triggered partly by military investment in the shadow of World War II, and partly by the long-established automobile industry (Rosenberg & Frischtak 1984).

But doubts remain. If the long waves do measurably exist, and if they are so regular, to explain them by exogenous forces requires either some regularly recurring mechanism in the outside socio-political process or an extraordinary set of historical coincidences. The 'exogenists' deny also the first, so they are left with the implausible second. The work of Forrester's global modelling team at MIT strongly suggests that the long wave, albeit a result of the complex interaction of the physical structure of the economy with the decision-making processes of individuals and firms, is indeed 'generated endogenously, and does not depend on random shocks such as gold discoveries to account for its persistence or turning points' (Sterman 1985, 128–9). In the MIT group's view it has two causes: firstly the fact that firms amplify unanticipated changes in demand, thus creating potential oscillations in the adjustment of productive capacity to demand; secondly, the fact that these tendencies are amplified through the linkages of firms to each other, to the labour market, and to the financial markets. The result is fluctuations in the demand for capital created by individual firms. So, rather than an innovation-based theory of long waves, the MIT researchers posit a theory of innovation based on long waves: midway through a capital expansion, based on commitment to an existing mix of technologies, opportunities for applying inventions are poor; but in the downturn the old capital depreciates, old bureaucracies are weakened, companies based on old technologies go bankrupt, and traditional methods cease to be sacrosanct (Sterman 1985).

Table 2.1 The four Kondratieff waves: different chronologies.

	Starting dates			
	First	Second	Third	Fourth
Kondratieff 1926	1790	1844–51	1890–96	
de Wolff 1929	?	1849–50	1896	
Schumpeter 1939	1787	1842–43	1897–98	
Dupriez 1947–78	1789–92	1846–51	1895–96	1939–46
Rostow 1978	1790	1848	1896	1935
Mandel 1980	?	1845	1892	1939–48
van Duijn 1983	1782	1845	1892	1948

Source: van Duijn (1983).

The statistical critics, on the other hand, are still open to the charge that there is no precise agreement on the timing of the Kondratieff waves (Table 2.1) and – a point they themselves make constantly – no unambiguous, authoritative standard list of innovations. Indeed, one plausible interpretation of Schumpeter is that, to paraphrase Orwell, though all innovations are equal, some innovations are more equal than others: certain key ones – better, perhaps, families of innovations constituting a new system – have the unique capacity to develop whole complexes of new industries and with them new socio-economic systems.

This is the argument of Carlota Perez, who suggests that the capitalist system contains two main subsystems, one techno-economic, the other socio-institutional. The long waves constitute successive modes of development, 'shaped in response to a specific technological style understood as a kind of paradigm for the most efficient organization of production' (Perez 1983, 360). In each upswing, there is a 'harmonic complementarity' between this technological style, or paradigm, and the socio-institutional framework; in the downswing comes a structural crisis, which is 'not only a process of "creative destruction" or "abnormal liquidation" in the economic sphere, but also in the socio-institutional', compelling adjustment – often, only after great strains – in the latter (ibid.)

Thus, Perez argues, in the downturn of each wave a new key technological factor precipitates crisis and ushers in the next: low-cost, steam-powered transportation at the end of the first Kondratieff; low-cost steel at the end of the second; low-cost energy at the end of the third; low-cost microelectronics at the end of the fourth. And each time the result is a new dominant technological style in the subsequent upswing: assembly line 'Fordism', shaped by cheap energy, in the transition from the third to the fourth Kondratieff; flexible batch production of information-intensive products in the transition from the fourth to the fifth (Perez 1983).

To achieve this transition, Perez suggests, there must be a period of 'experimentation at all organizational levels of society, characterized by the proliferation of reassessments, proposed solutions and trial-and-error behaviour stimulated by the increasing gravity of the crisis . . . in the face of the weight of tradition, of established ideas, of vested interests and other inertial forces which actively oppose the required transformations' (ibid., 365). Thus the new paradigm first emerges many decades before its eventual triumph: 'Taylorism', and then 'Fordism', basis of the mass production paradigm, existed in the 1910s and 1920s but became dominant only after World War II when institutional innovations – Keynesian state management,

Table 2.2 The four Kondratieff waves.

	First	Second	Third	Fourth
Dates	1787–1845	1846–95	1896–1947	1948–2000 (?)
key innovations	power loom; puddling	Bessemer steel; steamship	alternating current; electric light; automobile	transistor; computer; CIT
key industries	cotton; iron	steel; machine tools; ships	cars; electrical engineering; chemicals	electronics; computers; communications; aerospace; producer services
industrial organization	small factories; *laissez-faire*	large factories; capital concentration; joint stock company	giant factories; 'Fordism': cartels; finance capital	mixture of large 'Fordist' and small factories (subcontract); multinationals
labour	machine minders	craft labour	deskilling	bipolar
geography	migration to towns (coalfields, ports)	growth of towns on coalfields	age of conurbations	suburbanization; deurbanization; new industrial regions
international	Britain, workshop of world	Germany, American competition; capital export	USA, German leadership; colonization	American hegemony; Japanese challenge; rise of NICs; new international division of labour
historical	European wars; early railways	opening of North America; global transport and communications	world wars; early mass consumption; Great Depression	cold war; space race; 'global village'; mass consumption
rôle of state	minimal army/police	early imperialism	advanced imperialism; science and education	welfare state; warfare state; organized R&D

Sources: Various.

massive educational expansion, consumer credit, acceptance of trade unions, a new management style appropriate to the giant corporation – at last provided the necessary socio-organizational framework. A typology of long waves according to this wider framework is tentatively sketched in Table 2.2; the suggested dates represent partly our judgement, partly plain compromise as between various sources.

Seven key questions

The only fair conclusion about this debate, in late 1986, is that there is no single, agreed, long-wave theory of technological change and economic development; instead, long waves form a perspective shared by many different workers who differ greatly in their concerns, methodologies, theoretical outlook and empirical foci. In particular, we can define seven key questions on which there is still no agreement.

(1) *Fixed versus variable cycles* Some view the long waves as 'hard, determined and disciplined' and so predictable (Marchetti 1985, cf. Mensch 1979). Others see them as varying in length from country to country and time to time (Mandel 1975, 1980, van Duijn 1983, Freeman *et al.* 1982, Freeman 1984, 1985, Perez 1983, 1985): in their view special factors influence the process of economic upswing; technological innovation is not in itself the determinant, but forms a necessary precondition for growth to take place.

(2) *Definitions of innovation* Some focus mainly or solely on technological product or process innovations, especially the first (Mensch 1979); others stress the broader fivefold Schumpeterian definition of innovation, including new markets and modes of organization. The first group tend also to argue for long waves of fixed length; the second believe that they vary.

(3) *Kinds of technological innovation* Even within the narrowly defined technological innovations, some researchers stress major individual advances in particular products; others distinguish different levels and kinds of innovation, ranging upwards from the merely incremental, through Mensch's 'radical' innovations, to changes in technology systems and finally to changes in the entire technico-economic paradigm. Table 2.3 attempts to set these distinctions out schematically. The general focus of the most recent literature is on families and clusters of related innovations forming new technology systems, some of which (the 'pervasive' clusters) have

Table 2.3 Four categories of innovation.

	1 Incremental	2 Radical	3 Systematic	4 Paradigm
timing	continuous; rate varies with market conditions	uneven	uneven (long wave)	Uneven (long wave): diffusion may take long time
origin	within production process; market-led	deliberate R&D; influenced by market prospects	deliberate R&D plus organizational innovation	Deliberate R&D plus organizational, macro-economic and political change
degree of technical change	minor	major (single sector)	major (multisectoral)	major (pervasive)
economic impacts	minimal	minor and localized	major; creation of linked new sectors	major; Schumpeterian 'gales of creative destruction'
sociopolitical impacts	nil	minimal	significant	definitive: New socio-economic systems
industrial examples	panchromatic film; carbon ribbon; colour TV; 128K RAM	jet engine; TV; convergent IT	electrical engineering; electronics	mass production; mass consumption; welfare state

Source: Author's reworking of schema indicated in Freeman *et al.* (1982), Freeman (1983, 1985).

economy-wide applications (Freeman *et al.* 1982, Freeman 1984, 1985; Perez 1983, 1985).

(4) *Timing and diffusion* Some focus on the initial timing of innovations (Mensch 1979); this, as seen, demands agreement on what constitutes a radical innovation. Others stress their swarming and diffusion, which is not automatic or immediate but may be delayed for years or even decades by obstacles: the lack of specialized skills or knowledge, or of managerial capacity; perhaps most important, the failure of other sectors of the economy to adopt and adjust to the new technologies.

(5) *The causal mechanisms* A central point of debate still concerns the precise causal mechanisms, especially at the lower turning point: Are they endogenous, resulting from demand-pull or similar economic forces, or are they exogenously driven by technological change? Put another way, does technological innovation merely respond to changes in demand (Sterman 1985)? Or do clusters of radical innovations propel the economy upwards? To some extent, it can be argued, this debate is metaphysical: most likely, the truth lies between these extremes. Thus key technological advances must play a rôle; but they can only do so when demand conditions are ripe and when the organizational, managerial and institutional environment is right.

(6) *Rôle of organizations and institutions* Thus, much recent research stresses the crucial rôle of managerial, organizational, institutional and social innovation, both for the initial genesis and the subsequent diffusion of the technological innovations. But these are so far less well specified than the technological advances; we have only elements of a theory. One is the growth in scale of industrial organization, though Schumpeter himself thought that this would inhibit and finally throttle innovation (Schumpeter 1942). Another, associated, factor is the growth of specialized producer services which provide crucial functions such as R&D, design, marketing, distribution and servicing (Singelmann 1978, Gershuny & Miles 1983, Chandler 1962, 1975, 1984, Mowery 1984). Another is the linkage between industrial and finance capital, and in particular the significance of venture capital. Yet another is the changing nature of labour organization from one long wave to another: for instance, the rise of 'Fordism' in the mass-production industries of the third Kondratieff, and its more mixed fortunes – strong in some areas, weakening in others – during the fourth. Finally, there is the growing internationalization of all forms of economic relationships.

(7) *The state's rôle* This, of course, has grown immeasurably since Schumpeter wrote – though he clearly predicted that it would (Schumpeter 1942). In particular, the state has come to play a critical role in initial product innovation: as a customer for state-of-the-art high-technology products, especially in the defence field; as the most important funder of technological research and education; and as guardian of high-technology industries through tariff, trade and other industrial policies. But, at least until recently, this fact has not been well embodied in the-long waves literature.

We can now sum up on the state of the long waves art in the mid-1980s: just as the debate of the 1940s was largely a critical reaction to Schumpeter, so the debate of the 1980s has represented a similar reaction to Mensch. Mensch stressed the impact of radical product innovations at fixed intervals, triggered by a fall in profits on earlier investments; his critics have stressed innovation swarming, a wider definition of innovation that includes managerial and organizational changes, variable rates of diffusion, and the rôle of the state; there is no real agreement on the central issue of whether innovation is endogenously or exogenously determined.

Our approach

We must now state our own working hypotheses. They must be tentative, for our objective is to try to throw light on some of these points in the course of a longitudinal empirical study.

We base our own approach on a composite of the positions represented in recent work by Freeman *et al.*, Perez and the Forrester MIT group (Freeman *et al.* 1982, Freeman 1984, Perez 1983, 1985, Sterman 1985). We think that clusters of key interrelated technologies, developing through backward and forward linkages, are the real triggers of long waves; therefore, the aim should be to try to identify them. We believe that they tend to come forward at points when the returns from existing investments are declining. We accept the theory that as a result there is likely to be a disjuncture between the optimal application of the new technologies and the prevailing socio-economic mode of organization; thus, the development of initial innovations, and still more their general diffusion, may well be delayed by years or even decades; only when the two are again brought into harmony will the long wave upswing begin. Further, since the changes in organization are crucial they may permit the 'rejuvenation' of industries based on older technologies, which may therefore go through more than one Kondratieff upswing. However, the pace of this resolution may vary from country to country and from region to region – an important point, not well developed in the literature.

We think that the nature of this wider transformation will inevitably vary in character from one Kondratieff to another. Particularly, we would expect that in the most recent Kondratieff transitions the state must play a central rôle, in a variety of ways: general macro-economic management; the organization of both general and technical education; the direct and indirect promotion of research and development; industrial, trade and tariff policies; and, especially in the transition to the fourth Kondratieff, military policies. We would also, however, expect that this, too, might vary significantly from one country to another.

Finally, on the most critical point of dispute, we tend to the hypothesis that the exogenous–endogenous debate has been wrongly defined. If the transition from one Kondratieff to the next requires not merely clusters of hardware innovations but transformations of the entire socio-economic framework, then in a sense the whole process is endogenous: the underlying mechanism is indeed the laws of motion of capital, represented by a falling rate of profit, which eventually must trigger not merely a set of technological innovations but also changes in the economic, social and political superstructure. But, a critical point, the process is not an easy or an automatic one: it occurs only after periods of crisis that may strain the entire system to its foundations, as the history of the 1930s and 1940s eloquently demonstrates. Some countries may fail in part, even in large part, to make the necessary sociopolitical adaptations; hence, the rise and decline of nations (Wiener 1981, Olson 1982).

To examine these working hypotheses we take an approach different from any other recent study. We deliberately do not enter the statistical thickets of innovation analysis; instead we concentrate on a narrow range of related technologies and their related high–technology industries, through several generations of development, which, we would argue, have increasingly been in the vanguard of product innovation. As we go on to explain in Chapter 3, we do this for several reasons. Firstly, because we want to trace the longitudinal evolution of this group of technologies, and their supplying industries, over a long period (at least two and a half Kondratieff waves) so as to identify the timing and location of the key innovations that engendered them and the ways in which one relates to another. In this way, perhaps, we can throw light on the thesis that long waves may be triggered not by new technologies but by the rejuvenation of old ones and that the patterns of growth are related to other dimensions of innovation referred to earlier. Secondly, because we want to examine the thesis that such bunches of interrelated technologies may constitute not merely a source of synergistic economic growth through backward and forward linkages but also an entirely new technological–economic paradigm, with fundamental effects on the direction as well as the rate of economic development. Such hypotheses cannot be caught and tested by macrostatistical analysis; they demand the treatment that Schumpeter deliberately chose for *Business cycles*, the detailed analysis of the course of historical change. This, accordingly, is the mode of research we have chosen to follow in the study that follows.

3
New IT in a long-wave perspective

Information Technology and its derivative, New Information Technology, are relatively new additions to the jargon of social scientists and the language of popular discourse. In the UK the critical year was 1982, the Department of Trade and Industry's 'IT Awareness' year. So it is small wonder that there is still uncertainty about definition. New IT, it is generally accepted, is a subset of the wider category high-technology (or new technology) industry, which also includes other technologies (nuclear, biotechnology) of little direct relevance to IT. But the definition of high-technology industry, too, has generated much debate: as criteria, some suggest R&D levels, some R&D personnel, some the concentration of scientific and technological personnel, others simply the rate of growth (Markusen *et al.* 1986, Hall *et al.* 1987).

With IT, there is another problem: the relevant literature ranges widely, with different starting points and perspectives. Some comes from the long-wave perspective reviewed in Chapter 2. Other work stresses the rôle of information as an economic resource and a source of new industries and occupations (Porat 1977, OECD 1981). This relates to other studies seeking to reclassify the nature of economic activities in the light of recent economic changes, notably the growth of some parts of the service sector (Singelmann 1978, OECD 1981, Gershuny & Miles 1983). Closely related in turn are studies of the rôle and the value of information in contemporary industrial society. Another group, sociologically based, focuses on information as a source of economic power. Straddling them all are the popular writers who foresee the transition to the 'information society', constituting one of the great economic revolutions of mankind's history (Toffler 1980, 1983, Stonier 1983). We share some of their views regarding the significance of the changes; but we have to criticize their technological determinism, their shallow understanding of the economic and political context, and their lack of historic sense. A long-wave perspective, we shall argue, provides the missing elements.

Part of the definitional problem is that the words are ambiguous. *Technology*, defined in the dictionary as 'the science of practical or industrial arts' or 'a practice of an applied science', can either mean merely physical hardware or can extend to include non-material new forms of knowledge. *Information* is even more variously defined as 'intelligence, news', 'thing told, knowledge' and 'transfer or communication of knowledge'. At the most general level, IT seems mainly to refer to devices, systems and techniques for the generation, storage, retrieval, handling, transmission and communication of information, and its interaction with human activities. The definition in the United Kingdom's 1980 ACARD report is representative:

> On the one hand, the term may be confined to the actual equipment used to collect, store, process, transmit and display information. On the other, it may encompass not only the equipment (and the software that controls it) but its interactions with human activities and the management systems necessary if the capabilities of the new developments are to be fully exploited. We follow the latter approach in this report, in line with the UNESCO definition of information technology as: "The scientific, technological and engineering disciplines and the management techniques used in information handling and processing; their application; computers and their interaction with men and machines; and associated social, economic and cultural matters." (ACARD 1980, 12)

ACARD accordingly defines IT as encompassing a broad range of industries: important sectors of the electronic components industry, especially microelectronics; much electronic equipment, notably computers and peripherals; the whole communications industry, including the broadcasting authorities and the Post Office (then incorporating British Telecom) and the users and suppliers of information of all kinds. It excludes the application of microelectronic technologies to improved manufacturing processes, as through automation (ACARD 1980).

IT: old, new and newest

Such definitions, however, essentially refer to *New* IT; strictly, the newest of all forms of IT. They ignore the fact that information technology has a long history. For thousands of years, since the first cave paintings and the invention of writing, humans have used tools and techniques to collect, generate and record data (Inose & Pierce 1984). Over that long period we can distinguish three main phases of information technology. Firstly, simple pictorial representation and written language, evolving eventually into printing; its basic elements were paper, writing instruments, ink and printing presses. Secondly, a late 19th-century and early 20th-century cluster of mechanical, electromechanical and early electronic technologies: the telephone, the typewriter, the gramophone, the camera, the tabulating machine, eventually the radio and the television. Thirdly, a cluster of microelectronic technologies that have emerged in the second half of the present century and that are often taken, by ACARD and others, to constitute New IT. They include computers, robots and other information-handling production equipment, and office equipment.

In our long–wave historical view, however, the obvious links between this group and some elements of the second – their common dependence on electrical and electronic technology, and their evolution through continuous historical lines of research – make it misleading to separate them. In this book, therefore, we define *New Information Technology* (NIT) as encompassing both the latter two groups. We use the term *Convergent Information Technology* (CIT) to refer to the newest advances of the 1970s and 1980s, whereby computers and telecommunications are integrated into a single system of information processing and exchange.

Our resulting definition, in terms of UK 1968 SIC categories (and with the nearest equivalent 1980 UK SIC and UN categories), is shown in Table 3.1. We should stress now that there are considerable difficulties in applying such categories back over long periods of time. The greatest is the constant reclassification of products and industries, particularly the separation of items previously aggregated with others; thus most national statistical systems do not give separate figures for electronics until the mid-1960s, and the resulting reclassifications often have a considerable element of the arbitrary about them (Freeman *et al.* 1982). Similarly, in the 1980s, software production and servicing are inadequately separated out (OECD 1985a).

Refining our hypotheses

Our reasons for choosing this group of industries should be self-evident. Our hypothesis, which we set out to demonstrate later, is

Table 3.1 Definitions of New IT, by SIC categories.

UK SIC 1968		UK SIC 1980		UN ISIC 1958		UN ISIC 1968	
338	office machinery	3301	office machinery	370	manufacture of electrical machinery	3825	manufacture of office, computer, accounting machinery
351	photographic and document copying equipment	3302	electronic data equipment	391	manufacture of professional and scientific equipment	3832	manufacture of radio, TV, commercial equipment
354	scientific and and industrial instruments	3433	alarms and signal equipment; optical goods	392	manufacture of photographic and scientific goods	3851	manufacture of professional and scientific equipment
363	telegraph and telephone equipment	3441	telegraph and telephone equipment			3852	manufacture of photographic and optical goods
364	radio and electronic components	3442	electrical instruments and control systems				
365	broadcasting and sound reproducing equipment	3443	radio and electronic capital goods				
366	electronic computers	3444	electronic components (non-active)				
367	radio and electronic capital goods	3452	gramophone records				
		3453	electronic active components				
		3454	electronic consumer goods				
		3710	measuring instruments				
		3732	optical precision instruments				
		3733	photographic and cinema equipment				

Source: Central Statistical Office; United Nations.

that they began to constitute a 'carrier', albeit a minor one, in the fourth Kondratieff and can be expected to play a dominant rôle in the coming fifth Kondratieff. We must underline what we mean by this term: it stems from our working assumptions, set out at the end of Chapter 2. We suggested there that a new cluster of technologies will arrive on the economic stage during the downswing of a long wave; their general introduction will, however, be delayed by the need for wider socio-economic and political changes, entailing major and even agonizing transformation in the structure of the society.

Specifically, the New IT industries appeared in two clusters. One, the mechanical and electromechanical technologies, arose from innovations in the downswing of the second Kondratieff and became one of the carriers of the third – secondary to cars and electrical goods, to be sure, but nevertheless significant, particularly in their impact on the organization of the economy. Like cars and electrical products they became typical mass-production industries of that era, characterized by the early application of and Fordist techniques on the shopfloor, but also by systematic in-house research and development on a large scale. The second, the specifically electronic technologies, first developed out of electrical research in these same R&D organizations, towards the end of the third Kondratieff upswing, which were immediately applied to create the consumer radio industry of the 1920s. The main thrust, however, came from innovations of the third Kondratieff downswing and were generally applied in the upswing of the fourth. They too were characterized by even greater R&D efforts, this time bolstered by massive state contracting in the defence sector. At first they were characterized by Fordist techniques inherited from the previous era, but as they increasingly created the possibility of more specialized and customized production they enabled the beginnings of a reversal of the organizational methods of the previous Kondratieff.

We are suggesting, therefore, that clusters of innovatory industries may play key rôles in more than one Kondratieff. The mechanical New IT industries were clearly among the carriers of the third Kondratieff: the electronic industries of the fourth. But the precise rôle of the latter, in particular, is more complex. Its first real impact, through the radio industry, came in the downswing of the third Kondratieff where it played a clearly countercyclical rôle, creating appreciable expansions of production and employment at a time when the general tendency was decidedly opposite. Thence, a cluster of derived electronic consumer products – television, FM radio, high-fidelity recording – played a key rôle as early carriers of the fourth Kondratieff upswing; the early computer industry, albeit restricted to very specialized low-volume high-price military and industrial markets, played another. But the major further impact came just in the downswing of the fourth Kondratieff: a range of consumer and producer products based on the microprocessor, ranging from the pocket calculator through specialized applications (clocks, telephones, cars) to the personal computer and the early robots, played a countercyclical rôle analogous to that of radio in the previous downswing. The beginnings were already evident of the convergence of computer and telecommunications technologies, which we expect to play a key rôle as carrier of the fifth Kondratieff upswing.

We would like to examine this 20th–century sequence in particular detail, because it suggests possible important modifications of the basic long-wave hypotheses we have derived from Freeman, Perez, Forrester *et al.* In particular, it suggests that innovation clusters may form chains that extend over a considerable part of a Kondratieff wave. Thus the diode and triode valves of 1906–12, basic components of the radio industry of the 1920s and 1930s, came indirectly from a long line of R&D that began with the incandescent light of 1879; the microprocessor of 1971, basic component of all the new electronic products of the 1970s and 1980s, came more directly from the line of R&D that began with the transistor of 1947. In both cases the innovation chain appears to extend over some thirty years; the applications span a further two decades, thus covering the best part of an entire Kondratieff. This appears to explain both why certain innovations, coming at the end of the chain, can create countercyclical growth, and why they should then bring forth further chain reactions, which help trigger the subsequent upswing. Such are the highly tentative assumptions that we want to examine further in the history that now follows.

The problems of data

In it, we have used a great variety of historical sources, both statistical and other. Necessarily, given the very wide historical and geographical sweep of the study, we have had to rely in many places on secondary sources, especially technological, industrial and company histories. For more recent years, we have also been able to draw on the industrial and technological press, trade association publications, seminar and conference proceedings. But for the most part this is a work of scholarly compilation rather than of original research. There are two exceptions. For the UK we have been able to develop a continuous run of employment statistics that go back to the early 20th century; for the leading OECD countries, we have developed a unique comparative series for the 1970s and early 1980s, with – for a small subset – a more extended series going back to the early 1960s. Despite our best efforts, we were completely unable to develop more extended comparative international figures; the problem is quite simply that they do not exist.

PART TWO

The mechanical age

New IT in the second Kondratieff, 1846–95

4

The birth of new IT: telegraph and telephone

New IT was born in the 19th century, in the form of two electrical communications technologies invented 40 years apart: the telegraph (1837) and telephone (1876). It was thus firmly a creation of the second Kondratieff, though the subsequent development and impact of the telephone would constitute a major source of economic advance during the third long wave – and, indirectly, a potent influence during the fourth. So the earliest New IT, a child of the 19th century, was to produce children and grandchildren, forming an extended family of technologies which has vigorously reproduced itself throughout the 20th century.

True, in this long development other technologies have played a crucial rôle; late 20th-century semiconductors owe as much to 19th-century third-wave chemical technology as they do to second-wave electrical technology; the switching and printing functions in today's IT derive from early 20th-century printing technology. But there can be no doubt that the dominant line of development, which we now begin to trace, runs from the telegraph and the telephone to the extended electronics family of the late 20th century. It is thus important not only in itself but as the basis for the whole subsequent history.

The telegraph and telephone were in fact the first commercially useful electric technologies, that is, the first electrical innovations. The first came four decades before the major electrical advances we shall describe in Chapter 5, the second immediately in front of them. Yet they were very different from what followed: they depended on 'weak current' as distinguished from the high-voltage current necessary to light or heat cities or to drive electrical machinery. As we shall see in the next chapter, at first the new 'strong current' found little or no application in information technology, which developed between 1880 and 1930 as if electrical power still did not exist. That, in part, may have occurred just because the earliest New IT – dependent on weak current – was still growing and diffusing rapidly in these decades. For that reason, the following account starts at the beginning of the second Kondratieff and continues into the first years of the third.

The first electrical innovations

It is not surprising that electricity played no rôle whatsoever in the Industrial Revolution, which ushered in the first Kondratieff. For the basic scientific discoveries were only just beginning. As an historian of electricity has put it:

> Until a moment ago in man's long history, electricity was wholly inconceivable. It was not merely that its vast modern usefulness had yet to be discovered. It was impossible for man to imagine something that was, yet was neither solid, liquid or gas; something having no weight, being invisible, occupying no space, moving at enormous speed; yet for all this, something that was a normal part of nature. (Canby 1962, 9)

True, even the ancients had observed natural manifestations like lightning, static electricity or magnetism. But for centuries there was little appreciation of their relationship, let alone an explanation. Although in many other scientific fields the first advances came in the 17th century, in electricity they had to await the 18th. Between 1700 and 1750, Gray discovered conduction; de Cisternay, Dufay and Nollet identified two distinct types of electricity, and discovered the law that like charges repel and unlike charges attract; von Kleist and van Musschenbroek discovered the Leyden Jar, the first electrical condenser; Franklin proposed the use of positive and negative designations; Watson instantaneously transmitted an electrical charge along a two-mile wire. In 1767, Joseph Priestley inferred that the attraction of electricity was subject to the same laws as those of gravitation and set out the basic laws of electrostatics; in the 1770s, Lesage installed the first primitive telegraph, and Alexander Volta invented the first real electrometer, as well as a plate form of condenser. By the time of the Industrial Revolution, in the 1780s, fundamental advances had been made – in the UK, France, the Netherlands and Germany, and in the American colonies. But they were not yet sufficient in themselves to form the basis for commercial exploitation.

Perhaps the key advance came in 1796 with Volta's creation of the 'Volta Pile', a battery that constituted the first source of direct electric current; though its current was weak compared with the improved Leyden Jars, it was self-charging and continuous. Its demonstration at the Royal Society in London, in 1800, triggered a new epoch in electrical development (Canby 1962). Now, there began 'a communal interchange among great minds and small . . . a veritable explosion of discovery', bringing the principles on which modern electrical science is based (ibid., 41–2). Volta's battery was subject to numerous improvements, including the precursors of the modern dry-cell and storage batteries, and remained the principal source of electrical power until a practical dynamo was developed in the 1860s. Its existence led Davy, Faraday and others to discover and explore electrochemistry, electrolysis and the arc lamp. It facilitated the discoveries and contributions of Ampère, Oersted, Arago and others in the field of electromagnetism.

These advances were systematized and brought together in the work of Joseph Henry in the USA and Michael Faraday in England who, independently, discovered the laws of induction in 1832. Faraday was the first to publish his discovery that just as electricity can produce magnetism, so also can magnetism produce electricity via motion, the basis of our modern electric motors and generators. Henry even went further than Faraday and discovered that a changing magnetic field around a conductor induces an opposing current in the same conductor, the principle of self-induction; he discovered the principle of the transformer, whereby the voltage of fluctuating currents may be stepped up or down via induction from one coil to another. Henry also found that a varying current with a continually changing magnetic force could induce a complementary image of itself: a principle underlying the discovery of alternating power in the 1880s, as well as the transmission of speech and music as varying currents, and many other electronic applications (Canby 1962). These fundamental advances had come by the mid-1830s. Their commercial exploitation, in some important applications, did not come for another half-century. But in one, the telegraph, it came almost immediately.

The birth of the telegraph

We have seen that Lesage had developed a primitive electric telegraph in Geneva as early as 1774; and in 1812 came a wire telegraph based on battery power. But the modern electric telegraph depended not merely on the battery as a source of direct power but also on the electromagnet of 1825, Morse's relay, and an effective system for coding messages. In 1831 Joseph Henry of Albany, New York, after discovering the laws governing the design of coils of electromagnets, produced signals over a mile-long circuit by causing a horseshoe electromagnet to swing a pivoted permanent magnet and strike a bell: the first electromagnetic telegraph. In 1833, Gauss and Weber in Germany constructed a two-wire electric telegraph. The receiver was a compass-type device for detecting and indicating the flows of current, with each character represented by a certain set of movements of the needle. In 1837, Cooke and Wheatstone in England patented a six-wire, five-needle apparatus that could be read visually. The system was first installed on the railway between Paddington and West Drayton in July 1839 and was extended to Slough in 1843. In the 1840s, Cooke and Wheatstone developed a cheaper and more efficient single needle instrument, using equal-duration left and right deflections of the needle to denote characters in a code closely resembling that of the eventually more succesful chief rival system, that of Morse.

It was also in 1837 that Morse registered a patent for his system, five years after his first sketchbook ideas about the use of electromagnets in telegraphy made during a voyage home from Europe. At that time Morse was a landscape and portrait painter who knew nothing about even the most basic principles of electricity; but he was able to harness the technical knowledge of his friends and combine it with his own invention of the dot-dash code to produce the most widely used wire telegraph system.

These early innovators – Henry (Princeton University, Albany and the Smithsonian Institute), Wheatstone (King's College, London), Morse (a successful artist and professor of the arts of design in the university of the City of New York) – were keen inventors who were often academics. When Morse first successfully demonstrated his system at New York University in 1837 before Professor Gale and Alfred Vail, 'they appeared to be the only ones in this country at that time who had any appreciation of the great commercial possibilities of the telegraph . . . no private capitalist could be found who believed that this invisible and elusive electric fluid could possibly be put to any commercial use' (MacLaren 1943, 45). Morse had to wait until 1843 before the necessary $30 000 finance for his first practical system, over 43 miles between Washington and Baltimore, was provided by Congress. Only after the success of this initial endeavour, financed by the state, did the swarming process and rapid growth on the basis of private capital and entrepreneurship begin in the American telegraph industry. But then, quite quickly, demand came from railroads, the press, business and financial services; business, rather than private citizens, provided the basis for the industry's growth, in contrast to the position in Europe (Du Boff 1980).

Morse was thus able to found his own firm, the Magnetic Telegraph Company. From the very early years local telegraphy companies began to merge, partly in order to facilitate message transfer between various local systems and to avert telegraph chaos (Canby 1962). In 1856 the Western Union Telegraph Company was formed, unifying many small local telegraph firms, on the basis of Morse patents. Although Morse was initially barred from obtaining British patents, he did succeed in introducing his register as a competitor to Wheatstone's needles in many European countries commencing with France in 1845.

These growth decades of the telegraph saw many important improvement innovations, which considerably helped its functioning, reliability and cost–performance ratios. Morse replaced his embosser and inker registers with the much simpler sounder, which allowed skilled operators to transcribe messages more quickly and allowed average operating speeds of 20–25 words per minute. The demand for private wire systems that required no knowledge of code was met by Wheatstone's direct reading system and other similar systems developed in France and Germany. There were also many improvements in the types of distribution lines and wires and their insulation and more efficient utilization; one of the important innovations here was the duplex which made it possible for one wire to provide separate eastbound and westbound paths for simultaneous transmission, and in 1874 Edison invented a double duplex, called a quadruplex.

The innovatory origins of the telegraph thus came independently and simultaneously in the USA and in the UK. That makes the early challenge of Germany particularly interesting. For 30 years later, as we shall see in Chapter 6, the German electrical industry became (together with the American) the UK's chief competitor on the international stage; and its origins lay at this time, when, as an East German historian has put it, the infant electrical industry consisted exclusively of information technology (Barr 1966). These origins were curious ones. Werner Siemens, born near Hannover in 1816, enrolled in 1834 as an army officer because he had no other means of obtaining the scientific education he desired. In 1846, still a lieutenant posted to the Artillery Workshops in Berlin, he came into contact with the electric telegraph. It was a subject of great interest to the Prussian General Staff, who had been operating a semaphore telegraph since 1832 (Siemens 1957a, Weiher & Goetzeler 1981).

To develop and improve the telegraph, in October 1847 Siemens set up in partnership with a technician of the Physical Society (Physische Gesellschaft) of Berlin, Johann Halske, in a combined workshop and house in the Kreuzberg area of Berlin, close to the Anhalt railway station. The location was perhaps significant because it was close to the headquarters of the War Ministry. Here, for a time, he pursued an extraordinary treble life as army officer, official of the Prussian Telegraph Commission, and entrepreneur, a situation that came to an end only when he clashed with the Commission (Siemens 1957a, Kocka 1969, von Peschke 1981). His early home business was based almost exclusively on demand from the Prussian army, railways and the telegraph (Baar 1966, Siemens 1957a); in the 1860s, they expanded into long-distance telegraph cable ventures (Siemens 1957a).

Over in the UK, for some years immediately following Cooke and Wheatstone's initial system of 1837 the use of the telegraph was almost entirely confined to railways. The first British company formed to undertake the business of transmitting telegrams, the English Telegraph Company, was incorporated in 1846. For a time its activities were confined to the construction and maintenance of railway telegraphs and for some years it was regarded as a commercial failure. By the early 1850s many improvements had been made in the operation of the telegraph, a cable was laid from Dover to Calais, and the telegraph industry began to take off and make progress in the UK: 'numerous companies were formed, and keen competition led to considerable extensions of wires and reductions of tariffs, with the effect that the volume of business increased enormously' (*Garcke's manual of electrical undertakings* 1896, 2). Between 1857 and 1867 the Electric Telegraph Company's number of messages transmitted increased fourfold from 881 000 to 3 352 000 whilst the average cost per message fell from 4*s.* 1*d.* to 2*s.* 1*d.* (20.5p to 10.5p)

Throughout the 1860s there was much public dissatisfaction in Britain over the fact that the larger towns, where competition for the provision of telegraph services had been most keen, had unduly benefited at the expense of the smaller towns where the service was relatively less profitable. Telegraphy was becoming increasingly indispensable for industrial and commercial affairs. In February 1870, in response to political pressures, the government took over the telegraph companies, made a considerable reduction in the rates per message and a uniform tariff of 1s. (5p) per 20 words was established; this tariff remained in force until 1885 when it was further reduced to ½*d.* (0.2p) per word.

The nationalization of the British telegraph service companies arose from the fact that the growth and exploitation of the telegraph by private companies in the UK 'was not an unmixed blessing, as they were primarily concerned with making profit and not with catering for the varied needs of the public' (Fleming & Brocklehurst 1925, 16). Within a few years of the founding of the industry there was considerable agitation for the state to acquire the whole telegraph system. One of the leaders of this campaign was the Edinburgh Chamber of Commerce and it was followed by many other chambers; and most sections of the public press actively participated in the campaign. In response, the government ordered an inquiry into the running of the service in 1865. It found, among other things, that about 2800 towns were served by telegraphic service, a very small number of places compared with the 10 000 served by the postal service. Eventually a Bill came before Parliament seeking to remedy the defects of the service and nationalize it. The main arguments in favour of the Bill were set out by a Mr Scudmore, who in introducing the Bill argued that:

> the charges made by the telegraph companies were too high and tended to check the growth of telegraph; there were frequent delays of messages; that many important districts were not provided with facilities; that in many cases the telegraph office was inconveniently remote from the centre of business and was open for too small a portion of the day; that there was little or no improvement so long as the working of the telegraphs was conducted by commercial companies, striving to earn dividends and engaged in wasteful competition with each other. . . (quoted in Fleming & Brocklehurst 1925, 241–2).

It was also argued that the growth of telegraphy in Belgium and Switzerland had been greatly stimulated by the state ownership and control of the service and the consequent adoption of low charges; it was also argued that the association of the telegraph with the Post Office would benefit the public at large and eventually provide a substantial revenue to the state. The Bill was passed into law and the telegraph service was nationalized in 1870 on the basis of a 20 years' purchase of the profits of the year ending 30 June, 1868. The anticipated increase in the volume of business from the reduction of rates was realized when the number of messages relayed rose from 6.5 million in 1869, to 10.0 million in 1870 and 20.0 million in 1875. At the time of nationalization, there were about 30 different telegraph companies in the UK (apart from the railway companies) engaged in the transmission of telegrams and several different systems were utilized. In 1870, the total number of telegraph offices in the UK, including railway stations transacting telegraph business, was 2500. By 1896 it had increased to 9926 (*Garcke's manual* 1896).

The state monopoly did not extend to the colonies or to the international submarine telegraph business, in both of which private enterprises were the major actors. From the 1850s the growth of submarine telegraphy was considerable, with the total number of submarine cables approximating 1400 in 1896 and their length amounting to 161 385 nautical miles; of these, 1021 cables of 18 578 nautical miles were owned by various governments and 369 cables of 142 806 nautical miles by companies, most of whom were incorporated in the UK (*Garcke's manual* 1896).

Despite the invention of the telephone in the late 1870s, the telegraph continued to grow as a means of communication up until World War I, although it increasingly lost out to the telephone, as the data in Table 4.1 indicate; indeed, it remained the main form of long-distance international communication until well into the 20th century.

As with other later information technology systems based on electricity and electronics, the electrical telegraph technology and industry in turn facilitated the growth and development of other new sets of economic activities (product innovations). Together with the development of the railways, the telegraph was an important facilitating force in the development of mass-circulation newspapers and other media; the 'penny press' was able to feature

Table 4.1 UK Post Office telegraph revenues and messages handled, 1856–1 and 1911–2.

Year		Number of messages	Gross revenue (£)
1855		1 017 529	n.a.
1868		5 718 989	n.a.
1870–1		9 850 177	801 262
1883–4		32 843 120	1 789 223
1895–6		78 839 610	2 879 794
1907–8:	telegraphs	n.a.	3 100 940
	(telephones	n.a.	1 383 180)
1911–2:	telegraphs	n.a.	3 147 705
	(telephones	n.a.	2 962 736)

Sources:
(1) *Garcke's manual of electrical undertakings* (1896, 1913–4).
(2) Fleming & Brocklehurst (1925).
Note: The government took over the British telegraph operating firms in 1870, so prior to this date the numbers of messages refers to those carried by all the telegraph firms in the UK.

domestic and foreign hot news in place of journalism's traditional fare of stale editorial comments and commentary on information and news received by mail. The press wire service, a growing information industry even to this day in its new digitized and often wireless form, originated in the UK in the work of Reuter, who from the late 1840s was supplementing incomplete telegraphic communications systems on the continent by using carrier pigeons to bridge the gaps; from the 1860s he was operating submarine cables owned either wholly or partly by his news agency. In the USA, Associated Press was established in 1848 to pool telegraphic expenses, and it set up the first inter-city leased-wire service in that country in 1875. The telegraph thus played an important rôle in lowering the cost of information relevant to business transactions by speeding the flow of business information, extending the effective spatial scale of markets, and thereby improving the efficiency of the market economy (Du Boff 1980).

The emerging cable industry

From the 1840s on, a major part of the emerging telegraph industry was involved in the development, production, laying and maintenance of cables, wires and insulators capable of conveying messages via small electrical charges, across land and sea. The success and growth of the industry was dependent upon a whole set of novel inventions, innovations and literally 'heroic' entrepreneurship on the part of inventors, engineers and teams of workmen, especially in the case of submarine cables (Fleming & Brocklehurst 1925). Even in this very first, and rather primitive electrical industry, successful industrial innovation was systemic in character, depending upon advances in many aspects of electrical technology as well as those in chemicals and mechanical engineering, a point neglected in many accounts which focus on single product inventions or innovations in the apparatus concerned with sending and receiving the messages. Although these secondary innovations in the wire systems of conveying, transmission and distribution of telegraphic messages might be less dramatic and exciting, they were fundamental to the growth of the telegraph industry – and subsequently, since they constituted the first 'information highways', of telecommunications generally.

The early telegraph innovators were able to overcome a wide range of novel problems in this field. Morse's first telegraph line used glass doorknobs as insulators, but both British and American firms soon designed improved insulators of glass and porcelain which were used and improved upon by the later telephone and electrical power industries. The early telegraph industry also had to pioneer methods of laying and insulating underground cables to protect them from electrical failure due to moisture. The development of submarine transmission systems, from the late 1840s, presented a truly enormous set of technical and engineering challenges; these concerned not only the design of wire cables and electric insulation material with adequate mechanical strength and ability to withstand the rubbing action against rocks caused by the tides but also the development of new techniques of cable-laying from ships, based upon many expensive trial and error learning experiences and dogged determination (ibid.).

Indeed the problems of large-scale insulated wire and cable construction, the development of adequate insulators, lightning arresters, fuses and other components, which confronted the early telegraph industry, were entirely novel ones, requiring innovative solutions. They thus fed into and facilitated the subsequent growth of telephone communications systems, because telegraph lines were basically similar to early telephone lines, and almost invariably the same facilities were used for both. And they even greatly facilitated the development of the heavy duty transmission systems of the electric power and lighting industries which emerged after 1880.

London was the major communications node and centre for financing most of the international submarine cable business. In large part this arose from its position as the centre and metropolis of the hegemonic international industrial, trading and military power of the time. From the start, British-based cable and wire making firms had a strong international position, although one of the main companies was Siemens of Germany, which concentrated its submarine cable activities on its British branch. The British telegraphic wire and cable industry exported a high proportion of its total output (Scott 1958, Byatt 1979).

This strong innovative performance carried over into the manufacture of wire and cable for the electric lighting and power industries which developed from the 1880s. The heavier currents involved in the latter required new kinds of insulators and conductors compared to those used for telegraphy. The two main innovating firms in this respect in the early 1890s were not the established telegraph cable makers; one was Callenders, a road materials firm, which developed a system by which the conductors were laid in wooden troughs and covered with bitumen; the other was the British Insulated Wire Company, which developed Ferranti's simple but ingenious idea of using paper as a dielectric that substantially reduced the cost of high voltage cables. Although some paper-insulated flexible cables had been used in the USA for telegraphs, the use of paper insulation for electric power cables was first made and patented by Ferranti in 1888 for use on rigid tubes. Atherton, who founded the British Insulated Wire Company in 1891, invited Ferranti to join the company board. Using a variety of US patents and Ferranti's patents, the firm produced 11 000 volt paper-insulated flexible cable that was relatively cheap and easy to lay (Byatt 1979).

The British cable and wire industry continued to have a strong international position down to World War I, with exports more than five times the level of imports between 1906 and 1913 (ibid.). Indeed, throughout the third Kondratieff the manufacture of cables remained 'one of the strongest and most consistently successful sections of the British electrical industry' (COIT 1928a, 302). Two noteworthy features were that it was subject to less technical change than the electrical machinery industries (Byatt 1979) and that it was generally dominated by a relatively small number of firms (COIT 1928a; Byatt 1979).

The invention of the telephone

The telephone, as already seen, shared with the earlier telegraph a basis in weak electrical current. This second early electrical communications and information technology industry, however, differed from the telegraph in a number of respects. Firstly, the telephone permits the transmission of speech and other articulate sounds across great distances rather than signals that have to be interpreted by skilled operators or printed on paper via expensive equipment; it is thus not only an enhanced form of human electrical communication over space but one which can be more easily and efficiently operated by anyone with speech and hearing abilities. Secondly, although it may be an exaggeration to claim that 'the telephone is much more the product of a single inventive mind than the telegraph' (Fleming & Brocklehurst 1925, 249), it appears that 'only a few hardy pioneers had sufficient courage or imagination to undertake the much more difficult task of transmitting articulate speech over great distances, but their progress was comparatively swift' (MacLaren 1943, 50–1).

Thirdly, unlike the telegraph, there was a gap of almost 40 years between the initial scientific discoveries and their practical combination and application by Bell in 1876 (Finn 1985). One possible reason is that 'the social need was not great enough' at the time (ibid.); another may be a lack of 'innovation push', in that there were very few experiments and attempts at developing telephony between the 1830s and the 1870s. Perhaps this was because of the sheer novelty of the improvement in human communications represented by the telegraph system; the multi-faceted challenges posed by the telegraph system, and the improvement of its apparatus, wires, cables, insulators and other components, concentrated scarce inventive and innovative resources within this particular technological paradigm for four decades.

The principle of the telephone is based, like that of the telegraph, on magnetic induction: the human voice or other sound vibrates the air, which in turn vibrates a diaphragm; the movement of the diaphragm produces a corresponding vibration in an electric current. Thus the essential scientific discoveries necessary for the telephone were the concept of sound as a vibration which was known since the beginning of the 19th century, and the knowledge that vibration could be transferred to solid bodies (which Faraday had demonstrated in 1831 by showing how the vibrations of a piece of steel or iron could be converted into electrical impulses). Among the few to experiment seriously with sound transmission before Bell was Philip Reis, a German schoolteacher, who in the 1860s developed a crude device for the transmission of music; it was however ineffective for communicating speech (MacLaren 1943). In 1874 Elisha Gray, who was the chief electrician of the Western Electric Company of Chicago, invented a liquid form of transmitter which corrected the main defects of Reis's system. When Gray applied for a patent on 14 February 1876 he found that Alexander Graham Bell had filed an application for practically the same device a few hours earlier on the same day. The interference suit that followed was decided in favour of Bell (ibid.).

It was no accident that Bell's Scots grandfather was a professor of elocution; his father was a specialist in teaching the deaf and dumb, and he himself was a doctor who – after emigrating, first to Canada, then to the USA – became professor of vocal physiology in Boston, with a school for deaf mutes and others with defective speech (Birdsall & Cipolla 1980). He became interested in the analysis of sound waves and in methods of transmitting sound electrically by means of tuning forks; he was originally seeking to use this method to transmit several telegraphic messages simultaneously over the same wire, a problem that interested several telegraphic inventors at the time. Seeing a 'mechanical ear' at the

Massachusetts Institute of Technology – an early example of the serendipitous effect of urban agglomeration in innovation – he found that, by utilizing Faraday's discovery of electromagnetic induction, a miniature generator could be produced from a vibrating magnet placed close to a coil of wire giving rise to electrical waves which could be reproduced at a receiving instrument at the end of a line.

In pursuing this line of work in 1875, Bell was becoming discouraged despite his progress, for he felt that he was without adequate means for carrying out his ideas to the point of commercial success; but, in the summer of that year, he was strongly encouraged to continue his work by the 80-year-old Joseph Henry (MacLaren 1943, Birdsall & Cipolla 1980). In that same year he gained financial support from two Boston capitalists, his father-in-law Gardiner Greene Hubbard and Thomas Sanders, signing an agreement with them to form the Bell Telephone Company (Dummer 1983). He filed his patent for a 'speaking telegraph' on 14 February 1876, was granted it on 7 March, and successfully transmitted speech three days later; he spent the rest of that year preparing it for public use (Bruce 1973, Tucker 1978, Reich 1985).

It led to a huge battle with the Western Union company, who, rightly seeing the new instrument as a threat to their profits, commissioned Thomas Edison, then an employee, to develop an improved version using the rival patents of Elisha Grey. Edison developed a much improved transmitter, the weakest part of Bell's system, and a new type of receiver by 1878. The suit was eventually settled out of court by a compromise, whereby Bell was to have use of Edison's transmitter on payment of 20% of its profits on the rental of telephone instruments to Western Union and each company was to confine itself to its own particular field of activities.

Meanwhile, commercialization of the Bell system had begun in 1877 and a Boston electrical manufacturer, Charles Williams, was selected to make the first instruments. It was also decided at that time that the instrument would not be sold to the public but would remain the property of the telephone company. By the end of June there were 230 telephones in regular use throughout the USA, and two months later there were 1300 in use (MacLaren 1943). Soon it became evident that although the telephone derived technologically from the telegraph it demanded an entirely different system of organization. The telegraph had immediately been adopted as a communication and safety device by the railways, and a public message service developed as an ancillary to that primary purpose; consequently, the railway station became the public telegraph office (Birdsall & Cipolla 1980). But with the telephone many subscribers had to be interconnected, necessitating the principle of the central exchange; the first, at New Haven in Connecticut, opened in January 1878 (Tucker 1978). This was soon solved at a local level: by March 1880 there were 138 exchanges in operation in the USA, with 30 000 subscribers; by 1887, only a decade after the commercial introduction of the telephone, there were 743 main and 444 branch exchanges connecting over 150 000 subscribers with about 146 000 miles of wire (Finn 1985). But development of the telephone for long-distance communication was severely limited by both technical and, outside the USA, political constraints. It is a story that properly belongs to the third Kondratieff and will be treated in Chapter 6.

After settlement of the Western Union suit and a number of subsequent actions – a process that took until 1893 – a number of independent companies emerged in the USA. Yet the Bell system held such a predominant position 'that a great unified system was built up in which the smaller companies either disappeared or were granted exchange privileges which did not interfere with unrestricted communication throughout the country' (MacLaren 1943, 59). In Europe, however, politics intervened. By 1885, when the USA had 140 000 subscribers and 800 interurban connections, the UK – next in the world league – had a mere 10 000 and 80 respectively; the USA probably had twice as many telephones as the rest of the world combined, thus establishing a lead that was to last for many decades (Tucker 1978).

The telephone in Europe

This was not entirely due to lack of technical innovation. In the UK, the first working telephones were exhibited at the British Association meeting in Plymouth in 1877. In the following year an important British contribution to telephony was made by Professor D.E. Hughes, who found that finely divided carbon was a particularly suitable material for a loose-contact transmitter and that under favourable conditions this type of transmitter could be used to magnify weak sounds; he called the instrument a microphone. Hughes said in his paper that he considered the principle of the microphone a discovery rather than an invention and he thought others should have the benefit of his discovery. Rather than patent it himself, he allowed at least three different inventors of instruments to follow his designs quite closely (MacLaren 1943). Edison was the first to design a transducer using granules of carbonized hard coal, still used in present day microphones (Dummer 1983).

From the start, the rival Bell and Edison interests were not content with developing and expanding their telephone systems in the urban centres of the USA; they actively engaged in the transfer of their technology and systems to the cities of Europe (Hughes 1983). Bell himself came to Europe in 1877; Edison's telephone patents were promoted in England by Gourand, an American who often represented American banks in London and had valuable contacts among British capitalists. From 1878 Edison employed the services of Grosvenor Lowrey, whose associates, members of the great banking and investment house of Drexel, Morgan & Company, had the financial resources, the foreign contacts and the organizational wherewithal to move technology across national boundaries (ibid.), especially for major new systems and products such as telephones and electric lighting schemes, where such large financial resources were of the essence.

In 1878 the Telephone Company Limited was formed in London to acquire and work Bell's patent; in 1879 the Edison Telephone Company of London was formed, with plans to open telephone exchanges; in 1880 the two companies were amalgamated to form the United Telephone Company. But under the 1869 Telegraph Act, the British Post Office enjoyed a monopoly of all systems and devices for electrical communication; and it opposed the opening of public telephone exchanges, on the grounds that they infringed its exclusive rights, winning a High Court judgment to this effect. Thereupon the Post Office granted the telephone company licences to operate on payment of 10% of the gross revenues of the company. These licences were originally for specified areas; the companies had no power to make trunk lines between one town and another, or to open public call offices. In 1884 these restrictions were abolished.

The London and Globe Telephone Company was formed in 1882 and given a licence to compete with the existing exchanges, but in 1884 it was absorbed into the United Telephone Company. In 1889 the United Telephone Company amalgamated with some of the subsidiary companies to form the National Telephone Company. By 1896 this company had 'absorbed most of the companies holding Post Office licences, and is now the only company in the United Kingdom carrying on telephonic exchange business' (*Garcke's manual* 1896, 5). But a change in the law in 1899 allowed local authorities to operate municipal systems; and as late as 1912, there were at least three such systems (*Garcke's manual* 1912).

In 1896 the National Telephone Company had 654 exchanges and

88 093 subscribers' lines, handling about 350 000 000 messages per annum (*Garcke's manual* 1896). By this time there were also arrangements between the Post Office and the telephone company for sending telephone messages over the telegraph or via the postal services as express letters. By 1906 there were 10.15 telephones per 1000 population in the UK compared with 10.2 in Germany, the only other country, apart from the more advanced USA, that could then be compared with the UK in telephone development (Fleming & Brocklehurst 1925). The National Telephone Company continued to hold the monopoly until 1911, when it was taken over by the Post Office.

On the continent of Europe both the Bell and Edison companies moved quickly to introduce their new telephone systems. Late in 1877 Bell went to Germany and succeeded in obtaining the adoption of his system by the German postal service; several hundred telephones were ordered from Siemens Halske (MacLaren 1943). But many rival systems were also being developed in Europe at the time. Gower, a former Bell employee, developed one which met with considerable success in France and England. His enterprise became locked into a complicated set of patent suits with Bell and Edison companies and in 1880 the three interests united to form the Société Génerale des Téléphones in France and the United Telephone Company in the UK. Among the many other European devices were the Ader transmitter, which was quite popular in France, and the Crossley instrument; both of these and Gower's device were closely modelled on Hughes' discoveries (ibid.).

Many other competing systems and standards arose in Europe; 'The result of having so many companies starting up in a region where the field of activity was limited by national boundaries greatly retarded the creation of a universal intercommunicating system'; this was in sharp contrast to the USA, where a unified system was developed early under the Bell monopoly (ibid., 59). There, government supervision safeguarded the public against private monopoly power (ibid.); in many European countries the problem was later solved by direct government control.

Conclusions

The telegraph and telephone were thus not merely the earliest New IT industries but the earliest electrical industries to emerge. This may be due in large part to the fact that they required relatively low supplies of electricity compared to those required for electrical motors or lighting. Yet the telegraph industry emerged almost 40 years before the telephone, although the latter did not require any additional basic scientific discoveries for its development. The reasons for this time lag may be that the initial development of the telegraph itself constituted an extraordinary revolution in human communications potential; its systematic development and improvement may have presented a sufficient challenge to the prevailing technological (inventive and engineering) infrastructure. The telegraph's sheer novelty and revolutionary character is underlined by the fact that even in the heyday of *laissez-faire* capitalism, and even despite a burst of inventive activity attesting to its practicality, its early development everywhere demanded some state backing. Forty years later the telephone did not appear to suffer from any such hesitancy on the part of private entrepreneurs – at least in its birthplace, the USA. In contrast, the slow and relatively unsophisticated development in Britain led to demands for nationalization from the 1890s, finally realized in the 1911 Act (Clapham 1938).

One striking aspect of the development of the telegraph was the international character of the original scientific discoveries and resulting inventive activity, coupled with a level of international information exchange that is surprising in an era of poor transportation and communication facilities. Although Cooke and Wheatstone in the UK, and Siemens in Germany, developed a practical system at the same time as Morse in the USA, the latter is generally credited as the major founding father and innovator; even though his electrical knowledge was quite limited, his elegant solution to the problem of coding was a major contribution to the success of his system and its eventual dominance. In contrast, the telephone emerged from a relatively small group of inventors, and the credit for its initial innovation and development clearly lies with the work of Bell in the USA; Edison's pioneering development of an improved transmitter was also noteworthy, not only because it overcame the weak link in Bell's system, but also because the subsequent amalgamation of Bell and Edison telephone technologies ensured the rapid growth and dominance of this particular telephone technology both in the USA and internationally.

This distinction is important. At the onset of the second Kondratieff, in the mid-1830s, both inventive and innovative activity was finely balanced between the UK and the USA, with Germany in third place. Towards its end, in the mid-1870s, invention and innovation in New IT were firmly in American hands. As will emerge in Chapters 5 and 6, by this time the capacity for technological innovation generally was fading in the

UK; in the new electrical technologies of the period 1875–95 Germany and the USA shared the prize, whereas in the new mechanical information technologies the Americans took a clear world lead. One question to be asked, in Part Three, is how and why the locus of innovation should thus change.

PART THREE

The electrical age

New IT in the third Kondratieff, 1896–1947

5

The electrical revolution

The story of this chapter, in terms of the main theme of the book, will represent a large but a necessary diversion. In the 1870s and 1880s a series of major technological innovations brought not only the telephone (as already detailed in Ch. 4) but also, more fundamentally the development and harnessing of high-voltage electrical power for lighting, heating, and driving machinery. Together with equally important innovations in chemistry and the invention of the internal combustion engine, these technological advances provided the basis of the third Kondratieff long wave, which is conventionally dated as running from the late 1890s to the late 1940s. Yet, paradoxically, they were not immediately applied to the New IT. On the contrary: as will be shown in Chapter 6, most of the important advances that did occur at this time in New IT depended on mechanical technologies, and for several decades they developed as if electric power still did not exist.

Nevertheless, it is crucially important for our main theme that we outline at least the main features of the electrical revolution, for at least four main reasons. Firstly, the electrical industries, together with cars and chemicals, were the key high-technology industries of their day: they were for the third Kondratieff what electronics were for the fourth. Therefore a study of them can give critical comparative insights on the nature of the innovative process, and – of particular relevance for our main theme – its changing locus. Secondly, though not of immediate application, the new electrical technologies, and the firms that developed them, would eventually prove critical for the further development of New IT. This would occur especially through the early development of the electronics industries, particularly the commercialization of radio in the later decades of the third Kondratieff, 1920–50. Thirdly, from their earliest years the new electrical industries showed a clear tendency to concentration of capital and the formation of national monopolies based on patent rights. And this in turn would have important effects, both positive and negative, on the development of the electronics industries from the 1920s on.

Fourthly and finally, just because of the delay in reuniting electrical technology and information technology, the New IT industries of the 1880s and 1890s developed in large measure in a sheltered backwater, unsullied by the heavy currents of the main technological stream. Thus, as Chapter 6 will try to show, many major firms developed from a mechanical tradition, some few of which would later make a successful transition into the electronic age; others developed as electrical manufacturers, most of whom later made the necessary leap into electronics. The very end of the third Kondratieff, the 1930s and 1940s, was the era when the two technological streams, and their constituent major firms began to merge, setting up competitive cross-currents that would prove critically significant for the patterns of innovation in the following fourth Kondratieff upswing.

The basic innovations: batteries, generators, motors, transmission

For the electrical revolution to occur several different technological innovations must be brought together to form a system. Power must be generated. It must be harnessed to drive electrical motors. It must likewise be turned into a source of light. And, to be useful, it must be transmitted and distributed.

Once the voltaic battery had been demonstrated at the very end of the 18th century, a source of power existed. Thus it was possible to develop battery-powered electrical motors; the problem was that they were large, awkward and inefficient. One was Jacobi's, which powered a Russian paddle-wheel launch that could carry a dozen passengers silently at four miles an hour, sometimes for a whole day – though breakdowns were frequent and the huge acid-filled battery was especially dangerous in such a situation (Canby 1962). Also in the 1830s, Thomas Davenport drove a printing press and a miniature circular railroad with motors he devised himself. Although Davenport obtained a patent for his motor, it never became a commercial success. In France, during the 1840s, Froment was building sturdy motors to turn belts and run machinery.

The principles involved in the motor and generator were essentially the same. The motor is turned by magnetic force, exerted by reversible electromagnets converting electrical energy into motion through the intermediary of magnetic fields. The generator is essentially a motor turned backwards, in that it converts motion into electric energy as magnetic fields are forced across conductors. In the mid-19th century, attempts to produce electricity by crude generators using permanent magnets (magnetos) could only produce very weak currents due to the small power of the permanent magnets and the same inefficiencies that plagued the pioneering motors. After Faraday's contributions to electromagnetism in the early 1830s, three other Englishmen – Ritchie, Saxton and Clarke – all made important discoveries on the path to the development of a practical direct current generator, as did Pixii in France, Werner Siemens in Berlin and Pacinotti in Italy (MacLaren 1943). In 1862 a machine was developed which produced enough current to run an arc light; it powered several lighthouses of great brilliance by the standards of the day, as well as arc lights on board Prince Napoleon's yacht (Canby 1962).

In 1866, the dynamo, a generator that used electromagnets for the inducing field, was developed. In this year, six different dynamos were launched, each of which produced the necessary starting current for the electric field magnets. One, from Henry Wilde in Manchester, used a separate magneto generator to charge the field magnets; it drew particular attention for its high output and compact size. Wilde closed the reports on his experiments with an accurate prophecy: 'The transformation of mechanical energy into other modes of force on so large a scale and by so simple means will find new applications is one of my firm convictions . . . the electromagnet, as a source of electricity is destined hereafter to live in the lives of the millions of mankind when the memory of its origin, except with the curious and the learned, shall be forgotten' (quoted in MacLaren 1943, 113). Others who invented dynamos in that eventful year were Werner Siemens in Berlin, Varley and Wheatstone in England, and Moses Farmer in the USA.

In 1866, Varley, an electrician with the Atlantic Telegraph Company in London, applied for a patent for a self-excited generator having two field coils, and he received further patents over the succeeding years. In January 1867, Charles William Siemens took out a British patent for the same idea as Wilde's, based on the work of his brother, Werner, in Berlin (MacLaren 1943, Canby 1962). The latter immediately set about developing his invention and he fully grasped its practical significance. In a paper read to the Physical Society in Germany he said that it would make cheap and convenient electrical power available to industry everywhere that driving power was available (Siemens 1957a).

This spate of discoveries and inventions may be said to have completed the experimental and development stages of the design of direct-current generators; they 'had established the general principles and many details of design which lay at the foundation of development that foreshadowed the building up of a great industry' (MacLaren 1943, 114). Abundant electric power was at last on its way, but much additional work was needed for a fully practical and commercially viable generator, especially in perfecting mechanical details of design and in educating the public to the use of this new form of energy (MacLaren 1943).

Zenobe Theophile Gramme, born in Belgium in 1826, is generally credited with the invention of the definitive dynamo. He was living in Paris in 1870 when he took out his first patent for a direct-current generator which had a number of improvements on earlier devices. He actively published and promoted his discoveries, unlike some earlier inventors, and thus created such a demand that he was able to develop his business commercially (ibid.). His first direct current generators were used by the electroplating industry, but he also turned to the production of lighting generators and at the Vienna Electrical Exposition of 1873 he connected one of his big steam-driven dynamos to another, causing it to revolve at high speed as a motor. 'Suddenly it was clear that electricity could now do heavy work, transporting power through wires from place to place. It was a revelation – and immediately hundreds of minds turned to the possible uses of the idea' (Canby 1962, 83). To meet growing demand, Gramme brought out a full range of machines that could be used as either generators or motors. These machines had a wide sale for ten years or more in the face of severe competition. One was bought to illuminate Big Ben in London. Gramme can be credited with initiating the use of electric motors for industrial purposes and the commercial lighting of city streets by arc light in 1878 (MacLaren 1943).

The problem with his generators was one of efficiency. Their full-load efficiency was less than 50% compared to more than 90% today. Professors Thomson and Houston in Philadelphia in 1878, and William Siemens in Berlin at about the same time, pointed out the advantages of generator armatures having a lower resistance. In 1881 Siemens established the first public power station at Godalming in England (Landes 1969).

From 1880 the electrical machine became a commercial product; over the two closing decades of the 19th century manufacturing companies were established, which gradually developed into great organizations as individual effort became merged into corporate activity (MacLaren 1943). The growth of street lighting called for the development of special types of generators. Charles F. Brush of Cleveland, Ohio, is credited with the first truly successful street lighting system and in 1878 he designed two sizes of generators incorporating a regulator which made it possible to maintain a very steady current as lamps were thrown in or out of service and other improvements on Gramme's machines. After further improvements Brush started producing these lighting generators on a large scale after 1879; Brush improved the design of his machine again in the mid-1880s and, after his firm had merged into the General Electric Company in 1896, there were further improvements on the Brush design (ibid.).

In 1880, the Thomson-Houston Company was founded in New Britain, Connecticut, by the two Philadelphia professors who had designed a number of new generators using some principles similar to those of Brush, but with some strikingly original features; by 1883 the company had installed 22 arc-lighting plants operating a total of 1500 lights and by 1889 they shared with the Brush company the major part of the arc-lighting business of the USA (ibid.). Soon after this date, Thomson Houston absorbed the Brush company and began the manufacture of Brush generators at their factory at Lynn in Massachusetts.

During the late 1870s Thomas Edison in the USA was also developing a series of generators in connection with his incandescent lighting system. His first experimental generator was developed in 1879 and it included a number of important new features. After some improvements, Edison used these in 1881 and 1882 to launch 'his great electric lighting project by the installation of one hundred and fifty plants in which these generators were used for lighting theatres, hotels, residences and industrial shops' (ibid., 124). At the same time Edison was working on his much more ambitious plan of building the first general electrical distribution system from a central power station. In 1882 he opened the pioneer Pearl Street installation in New York, which established many of the principles for much of the electric light and power systems of today.

The growth of electric traction was based on the production of much larger generating units. One of the most important advances here was the Thomson-Houston Company's four-pole generator in 1890 which was rated at 100 horsepower and in the following year they brought out a 250 kW belted railway generator that was considered to be a great advance in producing larger capacities (MacLaren 1943). In the same year Westinghouse brought out their first railway generators, one of which was rated 125 horsepower and the other 250 horsepower; the following year it produced a six-pole generator rated at 500 horsepower. Between 1890 and 1892, the Edison company, Thomson-Houston and Sprague joined interests to form the General Electric Company (GEC). This new organization produced very large generators for the Intramural Railway, which were used during the Columbian Exhibition in Chicago in 1893. By the close of the century both Westinghouse and General Electric had developed designs for railway generators which were approaching the economic limit in capacity for this type of machine (ibid.).

By this time both GEC and Westinghouse had gathered together notable groups of engineers, scientists and technologists into their organizations, and from this time they had a dominating influence on the development of generators and other electrical apparatus in the USA (ibid.). By virtue of their large technical and engineering resources they were able to develop complete systems and solutions to the complex sets of novel problems involved in rendering the new inventions economically viable, safe and easy to use. They also had the resources to market and promote their products as well as educate and reassure their potential consumers of the reliability and safety of the strange new power source (ibid.). They had also developed links with financial and other organizations which ensured the early transfer of their technology abroad (BEAMA 1927, COIT 1928a, Hughes 1983).

The early direct-current generators and plants relied heavily upon huge storage batteries to smooth out the high fluctuations in the daily demand for electric power and to compensate for the breakdowns which were frequent in the early machines. In 1859–60 Plante in France made the most important advances in battery technology after Volta's pioneering work. He introduced the principle of the modern lead-plate battery, whereby lead plates are immersed in dilute sulphuric acid. However, when this battery was first brought out the generator had not yet been perfected and the charging current had to be obtained from a primary battery, so that under these conditions the storage battery could not yet be considered as an economical source of power (MacLaren 1943). In the early 1880s Camille Faure developed an improved method of forming the plates of a storage battery, and by this time generators could be used for charging and in combination with storage batteries. Faure used his battery in combination with an engine-driven generator to develop a continuously available source of power for lighting and small power applications; he also demonstrated that battery-operated cars might be used for passenger traffic on city streets. Indeed, in relation to that other major third long-wave new-technology industry, the automobile, it is important to note that for a time 'the battery-powered automobile was a more comfortable – and serious – rival to the noisy, smokey gasoline vehicles' (Canby 1962, 67). The early and rapid application of

Taylorist process and organizational innovations by Ford played a major rôle in the subsequent dominance of the latter form of automobile technology (Freeman *et al.* 1982).

There were other later improvements to the storage battery, one of the most notable being the Edison chloride cell. The most important applications of the storage batteries towards the end of the 19th century were made in the early central stations for direct-current generation. In one Edison plant, the Adams substation in Chicago, which was not untypical, a battery was installed which filled the entire building and was capable of supplying 22 400 A for eight hours continuously or much greater currents over shorter periods. By allowing the battery to take the fluctuations in the demand for electricity, the generators could operate under steady conditions, which tended to reduce costs in these early large direct current plants.

Until the end of the 19th century, the combined cost of generating equipment and batteries did not differ greatly from the cost of increased generating capacity that would be required to meet the varying demand if batteries were not used. But by 1900, following the great improvements that were made in both the cost and reliability of electric power generation, the economic balance tipped in favour of relying upon increased generating capacity in the newer plants (MacLaren 1943). The advances in the production of relatively cheap petrol powered, internal combustion engines in the early 1900s destroyed the earlier hopes of a very large application and demand in the operation of automobiles and trucks. But the electric storage battery was reduced to playing an important if restricted rôle as a secondary source of power in these products. In the early days of radio broadcasting, storage batteries were used for operating the direct-current sets, but again improvements eliminated this field of application. However, the production of batteries continues to be a significant if minor part of the overall electrical industrial sector to this day.

The early years of advances in electric power generation were focused on direct current, but this form of electricity lacked the critical attribute: the ability to transmit power at a distance (Canby 1962). When electricity was sent over distances of even a few miles its power fell off at an alarming rate, even though it was well understood from Gramme's time that the higher the voltage the smaller the loss. However, a major problem was that high voltage could not be used safely in crowded city areas, nor even controlled. Thus, until the early 1880s, most engineers concluded that electric power could never be more than locally organized. The solution lay in the long-neglected alternating current, which can induce a second current at any chosen voltage according to the ratio of turns between the primary coil of wire and the receiving or secondary coil. Not until 1880 did an Anglo-French team of John Gibbs and Lucien Gaulard develop and patent a 'series alternating system of distribution', based on their invention of the first true and practical alternating current transformer. In 1885 these patents were purchased by the American entrepreneur and manufacturer Westinghouse. Together with his chief engineer William Stanley and his firm's large team of R&D staff, Westinghouse set out to design a complete alternating power system to challenge the pioneering direct-power-based electric systems developed by Edison. This sparked off 'the great power fight' of the 1880s and 1890s between the proponents and interests involved in the two different systems, with Westinghouse leading the alternating current forces and Edison the direct current ones in the USA.

In 1886 the Westinghouse interests made a successful trial installation at Great Barrington, Massachusetts; the following year they completed the first commercial alternating current lighting system at Buffalo, New York. Electricity could now be generated at any convenient voltage, stepped up by transformers for economical transmission and again stepped down at substations for local distribution. Once the problems of high-tension safety had been solved it was possible to send electric power virtually any distance without significant loss of power.

Within a year of the successful Buffalo installation, Westinghouse had constructed 27 similar plants. The company concentrated its considerable technical resources in developing the necessary components for a complete alternating current distribution system. It set about the development of a line of generators, transformers, reliable meters, regulators and switchboards (MacLaren 1943). The growing success of alternating power was greatly enhanced by the discovery of the rotating magnetic field, the basis of practically all alternating current machinery today, and the invention of the alternating current motor by Nikola Tesla.

Born in Croatia (Yugoslavia) in 1856, Tesla worked for the Continental Edison Company in Paris in 1883 when he constructed his first induction motor in his own after-work leisure time. He sailed for the USA in 1884 and worked for a short time for Edison, but left after the two had a number of disagreements. He went on to perfect the polyphase alternating current system whereby an alternating current generator produces several interlocked currents which can lead a motor to receive a rotating magnetic field, with the fluctuations following each other around a circular ring of electromagnets. Tesla's system was the key to the heavy-duty industrial use of alternating current power. After leaving Edison, he sold his patents to Westinghouse, who quickly incorporated them into his rapidly improving alternating power generation and transmission systems. The first great heavy power installation on the polyphase principle was also the first major hydroelectric plant which was constructed by the Westinghouse enterprise and began operation at Niagara Falls in 1896. The big Niagara alternators produced some 15 000 horsepower, a massive leap forward on the 900 horsepower put out by the then pioneering Edison installation at Pearl Street, in 1882.

By the end of the 1890s it was clear that alternating current had won the battle of the systems; electric power could now be produced in great quantities and supplied over very large distances and had found a myriad of applications; it had become the basis of a major new set of industries. From the mid-1890s, 'the story of electric power thereafter was simply one of immense expansion along with the usual steady flow of technical improvement' (Canby 1962, 75–6). Although many improvement innovations were to follow, the broad contours of the modern electric technology system had been established.

The UK and the changing locus of innovation

Accounts of the development of inventions and innovations in electricity, like those in most other new technologies and industries, often convey conflicting pictures of the precise national origins of the key developments. This is often the result of unconscious bias and ethnocentricism on the part of commentators, and partly the result of the genuine difficulties that exist in any attempt to define what constitutes the key inventions and discoveries and the first true practical and commercial application of these. This is a particular problem in the field of electric and electronic technologies, where many of the major inventions and patents were the subject of complex and bitterly fought legal battles over the original authorship and ownership of patents; even the court decisions concerning many of these conflicts do not satisfy some commentators, who suggest that the technical issues were often beyond the grasp of the judges involved and that decisions often reflected nationalistic criteria or the sheer power of large enterprises over those with less resources at their disposal, rather than the more objective, technical merits of many cases.

One result is that English-language histories tend to give undue weight to American and British innovations and to downplay German and other contributions. Nevertheless, it is fairly clear that in the critical last two decades of the 19th century the lead in electrical innovation was taken by American and German entrepreneurs, with France and Britain playing a secondary rôle. This was despite the fact that the UK was a leading geographical locus of scientific discoveries and advances made earlier in the century and which laid out the principles upon which the technology and industry was built. However it was not only a failure in invention that was the British problem in the 1880s and 1890s; it was a failure in commercial innovation.

True, the UK did have some heroic entrepreneurs and innovators. Ferranti, often called the 'Edison of England', made a number of contributions for which he deserves to be regarded as a major innovator (Byatt 1979, Hughes 1983). In 1885–6, as a young man of 24, he completely redesigned and re-equipped the faulty Grosvenor Gallery scheme in London, developing a novel system based on transformers developed by the Ganz company of Budapest. He then went on to devise his famous Deptford scheme to supply high voltage electricity to central London at the then very high voltage of 10 000 watts and proposed the construction of 7500 kW capacity alternators, over ten times as big as anything constructed at the time. Although these giant machines were never finished, Ferranti started the Deptford scheme using 800 kW alternators which were still eight times the standard contemporary generator; in getting these into operation he had to overcome a number of novel technical problems which he did successfully, including making an important improvement in the design of steam pipes. His transmission system was also novel, using his own design for cables and impregnated brown paper as the dielectric. It was 'a work of genius, the first embodiment of the modern system of electricity supply using large generating stations in favourable sites and high-voltage transmission to consuming areas' (Byatt 1979, 102–3). But it was never completed and proved an over-ambitious failure, soon overtaken by rapid developments in polyphase alternating current (Hughes 1983).

Another important British innovation was the steam turbine of Charles Parsons (Byatt 1979). Parsons's first turbine of 1884 was a marked improvement over the power and efficiency offered by previous turbines; during the following decade he made many further improvements. The first Parsons turbines were installed in Newcastle in 1899 and Cambridge in 1892 by electric lighting companies formed by Parsons and his friends. The Westinghouse company bought the American rights for the Parsons turbine in

1895. As the power stations began to switch over to larger generating systems, the turbine came into its own and led to greatly increased steam economy (MacLaren 1943, Byatt 1979).

Lighting, traction and other electrical applications

The electrical revolution consisted not only in generating and transmitting power; it also consisted in finding and promoting new uses for it. These two sets of innovations, although treated separately here, should be seen as closely interwoven: developments in one stimulated and required advances in the other, as indicated by the key rôle played by the same leading innovative entrepreneurs and firms in both. Indeed, this growing innovative rôle of a few key large industrial organizations lay precisely in their capacity to combine and exploit a wide range of innovations to produce total new technology systems for the potential user, and to promote, package and market them non-technically, highlighting their potential at major exhibitions, and educating the public as to the safety of this strange, new source of power. Notice, for instance, how a leading figure like Edison paid great attention to the publicizing, packaging and delivering of new systems to the non-technical user in a user-friendly form, for example by making his lighting systems as identical as possible to the more familiar and established gas systems (MacLaren 1943).

One of the earliest and most obvious uses of electric power was in electric lighting. As early as 1801 Sir Humphrey Davy had noted that a brilliant spark appeared when he broke contact between two carbon rods which were connected to the two poles of a battery; but he could not maintain a continuous arc, apparently because of the weakness of his power source. By 1808 Davy had successfully demonstrated the arc lamp and found that various materials could be heated to incandescence by the passing of an electric current. It was several decades, however, before the arc lamp could be more than a laboratory instrument. There were experiments in the use of arc lighting in Britain for lighthouses in the late 1850s, but these remained much more expensive than oil (Byatt 1979). Lighthouses were an obvious use of arc lighting because of its powerful, concentrated illumination. In 1863 the Alliance Company of Paris made the first commercial applications in a number of lighthouses on the French coast.

A decade later, the introduction of the Gramme generator led to experiments in the streets of Paris and in Big Ben in London. In

1878 Paul Jablochkoff, a Russian engineer living in Paris, devised a new arc-lighting system and used an alternating current generator to provide the first demonstration of the use of electricity for street lighting (MacLaren 1943). Jablochkoff's system had a number of defects and was more expensive than gas lighting, but it stimulated great interest. In 1878 Siemens in Germany developed a successful arc light.

Already, the previous year, Charles F. Brush had developed a single light generator and arc lamp in Cleveland, Ohio. Over the next few years Brush rapidly developed his system: first installed in New York in 1880, within the next two or three years it was introduced into most of the large cities of the USA and many in Europe (MacLaren 1943). Brush's lamps were 1000 candlepower, in between the high power of the Siemens lamp and low power of Jablochkoff's, and overall his system was cheaper and more robust than that of his competitors. In 1880 the Anglo-American Brush Electric Lighting Corporation was formed to exploit the British rights to the Brush patents. There were many other competing inventors and innovators in the field of arc lighting from the late 1870s, and further improvements were made, especially in lengthening the life of the carbons; one of the most important improvements was that by Professor Lionel Marks of Harvard University, who succeeded in developing a successful way of sealing the carbon element in 1893 and who developed a high-voltage lamp in the following year.

Electric arc lighting opened up a new era of night illumination, and it was soon widely used in street lighting, illuminating large stores, public buildings, factories, railway yards and so on. But it was entirely unsuitable for domestic lighting, which awaited the development of a successful incandescent lamp. Throughout the 1870s there were a number of successful experiments in developing means of enclosing an electrical element in a vacuum tube so that when it was heated to incandescence it did not gradually waste away through oxidization by contact with the air (ibid.). In the late 1870s Sir Joseph Wilson Swan was stimulated by some of these advances to return to the development of the incandescent lamp, having himself carried out some pioneering work in 1860. In 1878 he originated the all-glass, hermetically sealed bulb, which later became universally adopted. By 1880 he was convinced that his lamp was a commercial proposition; he obtained a British patent and in the following year he founded the Swan Electric Light Company to develop and produce his system. The first installations of his system were in private residences, but in October 1881 he illuminated the Savoy Theatre in London with 1200 incandescent lights and with this accomplishment the success of his system was assured.

Meanwhile, however, a number of other inventors were working on an incandescent lamp and lighting system, principally in the UK and the USA. The most notable figure was the ubiquitous Edison, who took out a number of patents in 1878 and 1879. The success of Edison and Swan led a number of other manufacturers to enter the field and many new lights were introduced from the early 1880s. These resulted in several disputes regarding the priority of patents in both the UK and the USA; and in 1883 the British Edison and Swan interests merged after a clash over patents, and the combined operation used its strong patent position to try and eliminate competition in the British market (Byatt 1979). In the USA, Edison dominated the field, not only because of the technical qualities of his lamps but because of the attention he gave to the development of a complete system of illumination and his considerable skills in publicity (MacLaren 1943). In the UK, despite the important initial innovation of Swan, when the Edison–Swan monopoly ended in 1893 the electric lamp industry was dominated by foreign companies and there were significant imports up to the time of World War I (Byatt 1979). Between 1890 and 1914 there were no significant British contributions to improvements in electric lamps (ibid.).

Tramways proved to be the first major application of electricity for power. Early experiments in electric traction were made soon after the invention of improved dynamos. In 1879 Siemens and Halske introduced an electric railway at the Berlin Trade Fair; a year later, Edison made experiments in Menlo Park. In 1883 William Siemens built the electric tramway along the Giants' Causeway in Northern Ireland, covering a distance of eight miles. The first electric tramway on the British mainland was built on the beach at Brighton, using equipment from Siemens. These early electric trams used a third rail to distribute the electricity along the track; there were many experiments in using the newly invented accumulator as a source of power, but these were not generally successful (Fleming & Brocklehurst 1925, Byatt 1979).

In the late 1880s the USA provided the basic innovations for a safe and reliable urban electric tram system; they were the overhead trolley method of conveying power from a central power station to the vehicle and a robust traction motor (Passer 1953, Byatt 1979). By 1892 the major manufacturing companies were producing traction equipment in the USA on a large scale, and 31% of the country's tramway mileage was electrically powered. By 1897, 88% of the American tramway mileage was working electrically whereas only 9% of the British system had been converted. By that date Germany had also surged ahead of the UK in the electrification of the tramways, with the German manufacturers using licenses for the American systems or devising their own. The first electric overhead trolley in the UK was commissioned by the South Staffordshire Tramway in 1892, but 'little happened in Britain until the Americans arrived in force in the middle nineties' (Byatt 1979, 33). British Thomson-Houston was formed, with finance from Germany, but staffed by American engineers and using American equipment. After this, electric tramway development proceeded rapidly but existing British electrical companies made no great initiatives or efforts to promote electric traction in the early 1890s (Byatt 1979).

The first electric railway to be built in the UK (apart from light railways like Volk's in Brighton in 1884, or Blackpool-Fleetwood in 1885) was the City and South London Railway. This was a tube railway, installed by Mather and Platt and Siemens Brothers, and opened for traffic in 1890. The next 20 years witnessed a massive electrification and expansion of the London underground rail system, largely on the basis of American manufacturers' innovations and equipment and indeed American finance (ibid.). Over the same period a number of suburban railways and short inter-city railways were electrified.

The other major application of electric power during these early decades of the third Kondratieff was in factories. The early factory motors used direct current, but these had severe limitations. It was only from the second half of the 1880s, after Tesla's invention of the polyphase alternating current devices, that reliable, efficient and potentially cheap factory motors became possible. The polyphase system and induction motor were developed rapidly by American and European manufacturers from the late 1880s, but British electrical engineers played no part in it (ibid.). In fact British electrical engineers and manufacturers, compared to their USA and continental counterparts, largely neglected the potential of the induction motor until the turn of the century (Hughes 1983). Unlike the case with traction equipment, there was no large scale invasion of foreign firms in factory power equipment in the UK, and most imported equipment came from the Continent rather than the USA.

Soon, however, the diffusion and use of electric motors in factories became more important than those for traction. By 1907 factory motors accounted for about half of the total amount of electricity used in Britain; by 1912 they used three times as much power as did traction. Electricity provided only about 10% of all power used in British industry, but by 1912 it had risen to 25% and by 1924 the figure was about 25% (Byatt 1979). However, for most of the third long-wave upswing the application of electricity in British industry lagged behind that of Germany and the USA. As we shall see later, this was in part because of the small scale of enterprise. This retardation meant that the 'symbiotic relationship between industrialization and electrification', that could be found in Germany and the USA, was not present in British cities (Hughes 1983).

By 1914, then, this set of innovations constituted the major means by which electricity was being applied and harnessed. A number of other uses had been developed, especially in the sphere of domestic appliances, but these did not amount to any great significance at this stage. The 1920s and 1930s saw a continuing series of improvement innovations in the field of supply and transmission as well as new products utilizing electric power both for intermediate and final consumer functions, all of which led to a continuing growth of the electrical industries. However, it was only in the succeeding fourth Kondratieff, after World War II, that we find the largest absolute growth in the scale of electrical industries.

The years between the two world wars witnessed a continuing series of improvement innovations in the field of electrical supply and distribution as well as in products harnessing and utilizing electric power for both intermediate and final consumer functions. All of them contributed to the continuing growth of the electrical industries throughout the third long-wave and into the era of the fourth long-wave. The significant series of innovations which occurred after the 1890s, and which fed this industrial innovation and growth, may be regarded largely as refinements and modifications within what was by then a rapidly diffusing technological system; in essence, the universal polyphase system which emerged from the 1890s (the third electric system or 'generation' of electrical technologies) established the basic technological parameters for all subsequent developments (Passer 1953, Hughes 1983); the boundaries of the new technological paradigm were largely fixed and established by the 1890s (Preston *et al.* 1985).

Creating a new industry

This huge burst of innovation provided the basis for a wide range of new industries, including the manufacture of both production equipment and consumer goods, and the production, transmission and supply of electricity. The annual sales of the electrical manufacturing industry increased in the USA from a negligible amount in 1875 to over US$100 million in 1900 (Passer 1953); in Germany in 1882, the total number of employees in the electrical industry was too small to be separately enumerated, but by 1895 there were 15 000 people engaged in the industry and by 1907 this had risen to about 50 000 (Pollard & Holmes 1972). Nor was this innovation-led industrial output and employment growth confined to Germany and the USA, the chief loci of the major initial innovations; the UK too shared, but at a rate and to an extent that lagged behind the leaders. That is a theme for Chapter 7.

6

Information technology in the third Kondratieff: the mechanical byway and the electronic reunion

The third Kondratieff was the era of electricity, chemicals and cars. It was also an age of major innovations in information technology. The paradox is that many of these innovations – in the last decades of the second Kondratieff and the first decades of the third, all but one – occurred on a strange byway of technological development. They took place in the 1870s and 1880s, simultaneously with the great breakthroughs in electrical generation, electrical power and electrical lighting that we have described in Chapter 5. Some of the most important happened in the same places, and even at the hands of the same people, as the advances in electrical technology. Yet at the outset, most remained stubbornly non-electrical. Only after the main innovatory wave was over, in the first decade of the 20th century, did electricity and information technology fuse into a new science of electronics – and then, only partially and cautiously.

Chapter 4 has already shown that the second Kondratieff of 1843–95 had produced one major innovation in telecommunications in the form of the electric telegraph, and – towards its end – another, derived from it, in the form of the telephone. This too utilized the principle of weak electric current to transmit messages over long distances; it was independent of the momentous advances in electric power that were occurring in the same decade. Almost simultaneously came the recording of the human voice, but it was achieved without benefit of electric current of any kind. Other developments, that in effect created the modern office – the typewriter, the mimeograph – were purely mechanical devices, which similarly required no electricity to operate. It was not until the development of wireless telegraphy, in the first years of the 20th century, that fundamental advances in electrical research were allied to telecommunications.

In this chapter we use a wide set of secondary sources to retell the story of this strange technological divorce and subsequent reunion. As in Chapter 5, our main purpose is to show where the innovations took place, who made them, and how they became the source of major new industries. In that way, we wish to approach the central question of this study: how far, and in what ways, technological innovation becomes the basis of the economic rise and decline of nations, regions and cities. The analysis in this chapter will focus on the broad international stage. In Chapter 7 we shall look in closer detail at the consequences for Britain; in Chapter 8 we shall try to show how, within the leading industrial nations, the process of innovation was concentrated in a few key cities.

Electromechanical IT

In the transition from the second to the third Kondratieff, between 1870 and 1900, the modern office, and modern office technology, were born. But in a paradoxical way: the new electrical innovations passed them by. True, two key office technologies, the telephone and the dictating machine, made some use of electricity. But the telephone did not need heavy electric power, and the dictating machine was but little used. Another, the duplicating machine, might be electrically powered – though it need not be and indeed often was not. The fourth and most important, the typewriter, made almost no use of electricity before World War II. Neither did the fifth, the calculating machine. The real exception, the tabulating–accounting machine, was a minor one, albeit in terms of its rôle in technological evolution an interesting one.

The Telephone

The telephone, as already seen in Chapter 4, was an innovation of the 1870s. It was one of the building blocks of the third Kondratieff. Yet it occupies an anomalous position in the history of technology. For, as seen earlier, it was essentially a linear development of the electric telegraph, a classic second Kondratieff innovation; like it, it used weak power and did not at all depend – at any rate in early years – on the revolutionary advances in power generation and transmission made in the USA and in Germany between 1875 and 1890.

Chapter 4 has shown that once the telephone came it was almost immediately adopted: the transition from invention to innovation was virtually instantaneous. The new industry therefore began to show mushroom growth – above all in the USA – during the last years of the second Kondratieff, thus filling an interesting countercyclical niche that was later to be occupied by radio at the end of the third Kondratieff and by the personal computer at the end of the fourth. But until at least the first years of the subsequent fourth Kondratieff upswing its diffusion was severely limited by the available technology: interference, through induction from adjacent telegraph wires, made communication impossible at distances over 30 miles (50 km) when the telegraph was in use. The remedy, metallic circuits, was used experimentally in the UK in 1881 and in the USA in 1882 but did not become general until 1900 (Tucker 1978). Nevertheless, long-distance lines spread, most notably in the USA, where New York and Chicago were connected by 1892 and New York and San Francisco by 1915 (ibid.).

Distances of this latter order, however, required amplification through the earliest significant electronic innovation: the triode valve, developed simultaneously in 1906 by the American Lee De Forest and the German R. von Lieden (Tillman & Tucker 1978, Strandh 1979); not until 1912 did work by H.J. van der Bijl and F. Lowenstein lead to its use as an amplifier, notably on the New York–San Francisco line which opened in 1915 (Tillman & Tucker 1978, Brooks 1976, Fagen 1975, Reich 1985). The first long-distance cable in Europe using triode repeaters, over 300 miles (500 km) between Stockholm and Göteborg in Sweden, opened only in 1923 (Strandh 1979). Thus, though the telephone could develop up to a point without electronics, its true worldwide diffusion and interconnection had to await the start of the electronic age.

In Europe, however, there had been other obstacles. The state-owned post offices, which controlled the telegraphs, at first resisted the new competitor. Thus by 1885 the USA already had 140 000 subscribers and 800 interurban connections; the UK, next in the world league, had 10 000 and 80 respectively; at that time, the USA probably had twice as many phones as the rest of the world combined, thus establishing a lead which lasted for many decades (Tucker 1978). Within Europe, the Stockholm Telephone Company – thanks to the early innovative work of the entrepreneur Lars Magnus Ericsson – boasted more instruments than in any other European capital city (Strandh 1979).

In the UK, after a major battle, in 1896 the Post Office was given a monopoly of long-distance traffic and in 1912 (the municipal service in the city of Hull excepted) of all traffic; and similar arrangements came to prevail throughout Europe. Conservatively minded post offices might hold up the spread of the telephone for a while; but not for long. Everywhere during the 1880s and 1890s the trend was for a doubling in the number of subscribers every 2–3 years (Tucker 1978). Businesses, in particular, had to connect in order to compete.

Data before World War I are extremely sparse. The Census of Population recorded a total of 9442 employed in telegraph and telephone services in England and Wales in 1881; this had risen to 14 955 by 1891, 22 819 in 1901 and 27 431 in 1911. In the 1907 Census of Production, the National Telephone Company employed 6049 wage earners and 979 salaried staff, a total of 7028 employees. Thence, employment in the telegraph and telephone equipment manufacturing industry grew quite rapidly until the mid 1920s, after which it remained level for a decade; in the mid 1930s it again started to grow quite steeply. At the 1907 Census of Production, the industry's share of total national production remained surprisingly small: telegraph and telephone accessories accounted for a mere 2.7% of electrical manufacturing output, telephone and telegraph cables for 13.6%; the telephone industry accounted for only about 5% of total investment in electrical undertakings (Garcke 1907). Year-by-year investment figures show particular volatility both for telegraph and telephone, probably because of the financing of major international cable ventures (Garcke 1896).

The Typewriter

The typewriter too came in the 1870s, that golden decade of New IT innovation. It was, and for long remained, a quintessentially mechanical device for writing characters similar to those made by printers' type; and, when no more than a few copies were required, it immediately demonstrated its economic advantages over typeset printing. It also has many advantages over handwriting, including clarity and speed. The first recorded attempt to invent a typewriter occurred as early as 1714 when a British patent was registered for a machine whose description is similar to that of a typewriter, but no drawings of the device or information on its construction exist. In 1829, a machine called a typographer was patented in the USA; it consisted of type mounted on a semicircular frame which was turned to bring the required letter to a printing point and this was then depressed against the paper by means of a lever. The first recognizable modern machine, with separate type bars for each letter, came from a Marseille printer, Xavier Progin, in 1833 (Berry 1958). But nearly all these early machines were very large, and slower than handwriting (Anon. 1985). In 1867, the American inventor Christopher Latham Scholes (1819–90), a printer from Milwaukee, assisted by Clarence Glidden, constructed the first practical typewriter. Backed by a local entrepreneur, James Densmore, in the following six years they developed no less than 30 patented improved models that eventually wrote much faster than a pen (Beeching 1974, Birdsall & Cipolla 1980).

In 1873 Scholes signed a contract with the firm of E. Remington & Sons of Ilion, a small town between Albany and Syracuse in New York State, whose gun-making business was languishing with the end of the Civil War and a lull in the Indian campaigns, and who were already diversifying into sewing machines and agricultural equipment. The first typewriter was placed on the market in 1874 and the machine was renamed the Remington in 1876. It had such familiar features as the cylinder, line spacing and carriage return, the escapement for letter spacing by carriage movement, a 'qwerty' keyboard, and printing through an inked ribbon. The first shift-key typewriter, the Remington Model 2, appeared in 1878 (Beeching 1974, Birdsall & Cipolla 1980). After the Scholes machine was produced by Remington in 1873, typewriters were developed by a wide variety of American firms, eventually reaching more than 400 firms (Moran 1978). Mark Twain purchased one of the first Remingtons and soon after became the first author to submit a typewritten book manuscript.

This was symptomatic: in the land of its birth the typewriter, though resisted by the army of male copperplate-writing clerks, had an immediate impact on office work. From 1881, the American Young Women's Christian Association began to train women to type. Partly because of their work, there were already 60 000 women typists in the USA five years later; Rudyard Kipling, travelling there, wrote a letter describing the new phenomenon of the 'Typewriter Maiden' (Beeching 1974). Whatever employers' motives, the new office rapidly became a women's preserve, taking them out of the domestic service and factory jobs to which they had been confined; Scholes had predicted this, seeing it as a liberation (ibid.).

The diffusion of the new technology, characteristically, came a little later to the UK. And then it was marked by an invasion of the established American product: it was the name of Remington 'that England began to learn in the 'eighties, as she had learnt Singer's and Colt's names before' and in the typewriter field 'the other familiar names of the next twenty years, Yost and Blickensderfer and so forth, were also American' (Clapham 1938, 193). Following a familiar pattern for many new mechanical and electrical products at that time, American firms set up manufacturing plants in the UK. The import of typewriters and their parts for assembly in the UK grew from £383 000 in 1908, the first year of separate record, to £551 000 in 1913; by the outbreak of the war, typewriter imports into the UK 'were much more valuable than the imported American sewing machines, and their value was two thirds that of the imported agricultural machinery' (ibid., 193).

Whether in New York or London, the new offices might be lit from the start by electricity; after 1900, the female 'typewriters' might well travel to work by electric tramcar. But their typewriters remained mechanical. In 1933 improvements allowed the typewriter typeface to be changed and gave a uniform impression, allowing copy to be used directly for lithographic printing; in 1934 the carbon ribbon was introduced to give the type a better impression; in 1937 automatic justification enabled margins to be justified as in conventional typesetting (Moran 1978). But still, down to the mid-1930s, the obvious advantages of the electric typewriter – lighter touch, faster and more uniform typing, more legible and more numerous copies and less typist fatigue – went unrealized.

Perhaps the cause was economic: unaided female fingers could depress the keys fast and cheaply enough to make adoption of electric power an unviable proposition. But the main reason seems to have been technical. Although Thomas Edison had obtained a patent for an electric typewriter using a printing wheel – a primitive version of the IBM Selectric – as early as 1871, in association with George Arrington of Washington, it then lacked an adequate power source and never seems to have come to production (Beeching 1974, Conot 1979). The first production machine was the American Blickensderfer of 1902; it cannot have sold well, for no examples have survived (Beeching 1974). Although Remington introduced an electrical model in 1927, neither it nor any other company seems to have had much success with such machines down to World War II; they had too many faults (ibid.).

The breakthrough came with International Business Machines (IBM). In 1933 it acquired a small company, Electromatic Typewriters. Originally called the North-East Appliances Company and owned by the North-East Electric Company, it had become independent when its parent was taken over by General Motors in 1929. It had been producing an electric typewriter since 1924, but lacked capital and had been conspicuously unsuccessful; selling at nearly twice the price of the most prestigious manual, the

Underwood, it could not compete (Sobel 1983). IBM's President, Thomas J. Watson, immediately pumped $1 million into the development of the electric typewriter.

Put on the market in 1935, it was still selling only 6000 a year by the end of the decade, 5% of the total industry market (ibid.) Nevertheless, it was the first successful electric typewriter in the USA; and, stealing a total march on the well established manual typewriter companies, it marked the beginning of IBM's relentless march to dominance of the world's office machinery market. Thereafter, IBM's product innovations came at regular intervals of about a decade: proportional spacing – originally developed in 1941, but postponed by the war effort – came on the market in 1946; the Selectric, with the unique golf-ball, followed in 1961. Three years before that, IBM had already sold its millionth typewriter (Beeching 1974). Thus, ironically, the electric typewriter was perfected only after the electronic age was well advanced.

The dictation machine

The Victorian typist, and her successor a couple of generations later, invariably was transcribing material that had been recorded by nothing more technologically sophisticated than a pencil and pad. The one innovation she used – and that, in its timing, was quintessentially second Kondratieff – was the shorthand system of Sir Isaac Pitman (1813–97), as laid down in his *Stenographic sound-hand* of 1837. By 1889 an investigation showed that it, or some modification of it, was used by 97% of American shorthand writers (Russon 1985). From 1893 onward it suffered inroads from the rival system developed by the Irish-born John Robert Gregg (1867–1948) and later by the Speedwriting system of Emma Dearborn, an instructor at Columbia University (ibid.).

In comparison, the dictation machine, though developed almost simultaneously with the typewriter and the telephone, made little progress. The story is an odd one, often repeated in technological history, of how an innovation gets diverted into a different use. The idea of recording sound can be traced to 1856 when L. Scott described a 'phonautograph' consisting of a solid medium covered in lamp-black to record sound waves by means of lateral undulations of a line (Tucker 1978). In 1877, C. Cros in France described a system in which the recording was to be made by Scott's method on a disc with spiral traverse, from which a steel copy was to be made by photoetching; this could be used for reproduction by a stylus coupled to a diaphragm. In the same year, Thomas Alva Edison (1847–1931) announced his phonograph, which used metal foil fastened to a rotating cylinder to record sound waves by variations in the depth of the impression, the so-called 'hill and dale' method. Edison developed his device as a dictation machine capable of both recording and playback, with stylus and diaphragm serving both purposes; indeed, his original idea had been to use it to turn written messages into sounds, which would be transmitted by telephone and then recorded in order to turn them back into writing again – hence phonograph, 'sound writer'. Only as he worked on the problem did he see that the human voice might be recorded, and even later that the device might be used for recording other than from a telephone (Conot 1979, Chew 1981). Then he allowed the patent to lapse, only returning to the idea of using the machine for stenographic use in 1886 (Conot 1979).

But 'clerks were cheaper than office machinery' (Birdsall & Cipolla 1980) and the idea did not catch on. The main reason, apart from Luddism on the part of stenographers and the economic depression of the late 1880s, was that 'industry was simply not ready for the dictating machine' (Chew 1981, 13); 'for the average firm the use of the dictating machine was at first a luxury, then an alternative to normal stenography and only now is it becoming a necessity' (Chew 1981).

Ten years after the original phonograph, Berliner introduced the first flat-disc record and also a method for making many shellac copies. Berliner introduced the term 'gramophone' and the mass production of disc records from a permanent master with lateral undulation on a spiral groove. About the same time appeared the rival graphophone developed by Alexander Graham Bell's cousin Chichester and his collaborator Charles Sumner Tainter. This spurred Edison to develop the phonograph of 1895 as a pure home entertainment machine. That demanded the centralization of the recording process, a means for duplicating and mass producing records from a single master, and a massive simplification of the machine itself to put it within the reach of the average person (Chew 1981). The key to this last was a spring-driven machine, first developed by the Columbia Phonograph Company as early as 1892 and by Edison in 1894 (ibid.).

The paradox was intriguing. As one historian put it at the end of the 1950s, though sound recording and reproduction had for half a century been a part of electric communications, the first half-century of the art had practically no connection with electricity (Jarvis 1958, 1262). After Edison's invention of 1877, the machine remained essentially acoustic, making some limited technical progress. Emil Berliner, an American entrepreneur, in 1888 introduced his 'gramophone' and soon after developed the idea of mass production of discs from a permanent master, which created a large-scale entertainment industry (Jarvis 1958, 1262, Chew 1981). But electric recording, through a microphone and electronic amplifiers, developed only from 1926 (Jarvis 1958, Tucker 1978). Electromagnetic pickups, which convert the vibrations of the stylus into electrical signals for amplification and use with a loudspeaker, followed soon afterwards. Bell's graphophone of 1885 was treadle-driven; Edison's improved phonograph of 1888 was powered by an electric battery with a life of only 15 hours (Chew 1981); the same power pack was used in early dictation machines at about the same time (Beeching 1974). Still, even into the 1930s and 1940s, the typical gramophone or phonograph remained an acoustic instrument powered by a spring motor, entirely independent of electricity. The developments that changed this – crystal-type pickups, hi-fi systems, long-playing records, the application of the transistor and of integrated circuits, stereophonic recording and reproducing systems – all became general from the late 1940s to the late 1950s: that is, at the start of the fourth Kondratieff (Tucker 1958).

Electromagnetic recording of sound, on magnetized tape or wire, also came into general use only in the late 1940s, though it had a much longer history. The first practical system was introduced by Poulsen in Denmark around 1897, using a steel wire wound spirally on a drum, with a magnetic recording and reading head travelling longitudinally as the drum rotated. The sound to be recorded was converted by a microphone into an electrical current which, by flowing through a coil in the magnetic head, varied the magnetization of the wire and remained recorded until removed by demagnetization. This had the advantage that the magnetized material or tape could be used repeatedly by demagnetizing it and recording new material. Poulsen tried to exploit the invention commercially in the USA, and from 1903 his company produced dictating machines and telephone recorders; but the slow rewinding speeds, difficulty of playback and other technical obstacles, as well as managerial quarrels, put the company into receivership (Jewkes *et al.* 1958, Tucker 1978).

In the 1920s, the US Naval Research Laboratory made many improvements in magnetic recording for transmitting telegraph signals at high speed, which became important in the rapid growth of magnetic recording from the 1940s. The powerful electronic amplifiers of the 1920s also helped correct the worst failings of the early recording machines, but commercial interest and application remained slow. In the late 1920s in Germany, Kurt Stille set up a company to sell licences to manufacture magnetic recording equipment. Blattner, a motion-picture promoter, obtained a licence in the UK and sold rights to the British Marconi Company, which manufactured the Blattnerphone tape recorders used by the BBC. Karl Bauer, and later C. Lorenz, one of the largest German communications companies, bought licences and produced combined dictating and telephone recording machines (Jewkes *et al.* 1958).

A critical advance was plastic-tape media, coated with fine magnetic particles, to replace steel wire and tape; developed in Germany, from about 1937 it was used by AEG in their Magnetophon machine. Further development work took place in I.G. Farben and the German State Broadcasting Service during World War II. In the late 1930s and early 1940s several American firms produced improved tape recorders, and Marvin Camras of the Armour Research Foundation made many improvements in coating materials for tapes and in the use of magnetic sound for motion pictures (ibid.).

World War II, as in many other IT fields, provided a major impetus for the development of better magnetic recording devices (ibid.). After the war, the US Alien Property Custodian held the patents to many of the important German patents, which were made available for licensing to manufacturers. By then, high-quality recorders were being manufactured, mainly, it appears, by small companies (ibid.). Thus the large-scale application of electricity and electronic sound recording and reproduction, apart from its use in motion pictures in the late 1920s, came after World War II with the development of high-fidelity recording on tape (ibid.). In turn, these later facilitated the development of reliable and cheap recording devices for information other than music and speech, such as in modern computer storage systems.

Nevertheless, Edison's phonograph – non-electric as it was, largely spurned for office use – had a great significance in the history of information technology. For, as one biographer has written, it was 'the precursor of all modern forms of mechanical memory' (Wachhorst 1981, 22). And, by then in quick succession inventing the incandescent electric lamp (1879) and then discovering the so-called Edison effect (1882), Edison was in effect laying the foundations of modern electronics, though he did not know it. Later these technologies would come together to provide some of the crucial building blocks of new information technology; but this is a story of the fourth Kondratieff.

The duplicator

The third element in the modern office – not in all, but in some – was a duplicating machine to make multiple copies from a single master. The most typical, which survived in many offices until displaced by the photocopier at the end of the 1960s, was the stencil duplicator. The stencil itself was made by direct impression of the typewriter keys, cutting through the wax-treated paper; thence, it was inserted on to a roller to make a rotary impression. Like the typewriter itself, it was essentially an outgrowth of mechanical printing technology; like the typewriter, it could and often did remain hand-operated.

Like so many other pieces of the New IT, this invention came from Thomas Edison – and, as with the phonograph, as a by-product of a quite different search. In 1875 Edison developed an 'electric pen' capable of making copies of handwriting on to a stencil for reproduction through an 'autographic press'; it failed because of the lack of electric current and competition from the first typewriters. Edison accordingly developed a stencil typewriter for use with a mimeographic press, selling the patent to Western Electric who seem to have done nothing with it. In 1887 Edison bought back the rights and sold them to A.B. Dick, a midwestern lumber manufacturer, who registered the Mimeograph as a trade mark in 1889 and marketed it aggressively from the 1890s onwards (Chandler 1975, Conot 1979). As the typewriter diffused into offices, it logically followed behind.

The calculating machine

Commercial offices would also use a variety of calculating machines. They were built on the same principle as the very first, developed by the great French mathematician Blaise Pascal in 1642, and the first commercial version, developed – also in France – by Charles Xavier Thomas de Colmar in 1820: they used rotating wheels to make the elementary arithmetical calculations of addition, subtraction, multiplication and addition. De Colmar's version was in regular commercial use, under the trade name Arithometer, until the 1930s; its chief rival, invented by the American Frank Stephen Baldwin in 1872 and developed by Willgodt Teophil Odhner, was found in offices everywhere, in the USA as the Ohdner wheel, in Europe as the Brunswiga, until its displacement by the pocket calculator as late as the 1970s (Margerison 1978, Strandh 1979). All these machines depended on hand setting of a pointer to the required input values, followed by rotation of a handle; they needed no power assistance.

Keyboard machines (the Comptometer of 1887, developed in the USA by Dorr Eugene Fell, the business machine developed by William S. Burroughs which came on the market from 1890) were similarly at first hand-operated, though an electrically powered Burroughs was available from 1906 (Margerison 1978). By the 1930s the most sophisticated of these electric machines were fully automatic even for multiplication and division (ibid.); for these functions, however, they depended on repeated mechanical revolutions, and were necessarily somewhat slow. They borrowed printing features from the typewriter; the more sophisticated among them performed complete bookkeeping functions, thus acting as a natural bridge for the entry of typewriting into the data-processing field.

The tabulating–accounting machine

The most interesting of these third Kondratieff business machines was, however, a rather special case. In 1879 Herman Hollerith (1860–1929), born of German parents in Buffalo, New York, graduated from Columbia University's School of Mines and began work for the US Bureau of the Census. Inspired by a remark by Dr John Shaw Billings, in charge of vital statistics, after leaving the Bureau he continued to work on a system to mechanize the tedious job of tabulating the census results. His plan, patented in 1884–9 and adopted for processing the 1890 results, borrowed the principle of the Jacquard loom, developed in 1805 by the Frenchman Joseph Marie Jacquard (1752–1834) to automate the weaving of fine patterned fabrics (Margerison 1978, Strandh 1979, Sobel 1983). The principle of the Jacquard loom was a series of punched cards which controlled sensitive pins, which in turn animated the operations of the loom. Charles Babbage, the visionary 19th-century British computer pioneer, had also foreseen its adaptation for other purposes, thinking of the possibility of 'weaving algebraic patterns of the same beauty as a Jacquard fabric' (Strandh 1979). Hollerith added a principle derived from the electric telegraph: electrically sensitized pins would 'read' holes punched in cards made out of a non-conducting material and sort them accordingly, whereupon electromechanical counters would record the numbers of cards falling into each category (Margerison 1978, Strandh 1979).

The invention, originally battery powered, thus depended on a combination of weak-current technology and electric power. Eagerly adopted by census offices in many countries, it clearly had many other applications. In 1896 Hollerith set up the Tabulating

Machine Company, incorporated in New York, to exploit it; it was soon producing a variety of machines for punching, storing and tabulating data. In 1911, 'short of both cash and innovation' (Sobel 1983, 20), it merged with the Computing Scale Company of America, based in Dayton, Ohio, and the International Time Recording Company of New York, based in Endicott, New York (the first a manufacturer of weighing machines, the second of workplace time clocks) to form the Computing-Tabulating-Recording Company. Hollerith continued to work for this company as consulting engineer until 1921, three years before it was renamed International Business Machines (Starr 1944, Margerison 1978, Strandh 1979).

Thomas J. Watson, who came as president of the new company from National Cash Register (NCR) in 1914, rapidly built up the tabulating-accounting machine business as a mainstay of its operations. Its competitors included the old-established and somewhat conservative Detroit-based Burroughs Adding Machine Company, the even more conservative NCR, and the Powers Accounting Machine Company, founded by yet another ex-Census Bureau official, James Powers, which merged in 1927 with Remington Typewriters and others to form Remington Rand. Ironically, this company, merging in 1955 with Sperry Gyroscope to form Sperry Rand, then became IBM's major competitor in the burgeoning computer business (Margerison 1978, Sobel 1983). Under Watson IBM specialized in the leasing of large-scale, tailor-made systems (comprising punches, sorters and accounting machines) to larger companies; all were electrically powered from the start, but Powers and Burroughs began to switch only in the 1920s (Sobel 1983). Their great triumph was, in 1935, to win the contract to equip the new Social Security Administration under Roosevelt's new deal. Between 1928 and 1939 IBM overhauled Burroughs, NCR and Remington Rand to become the firm with by far the highest earnings of any business machine company (ibid.).

The link between IBM's business in the 1930s and its postwar triumph in computers is a fairly direct one; it is both technological and commercial. Technologically the computer needed an input device which the punched card, soon known to the world as the IBM card, was unusually well suited to provide. Its punched holes essentially provided the answers to yes–no questions and thus proved ideal for the binary logic developed by the English mathematician George Boole (1815–64) in his *Mathematical analysis of logic* (1847), which was rediscovered in the 1940s to provide the basis for all subsequent computer operations (Strandh 1979). Thus, by a strange set of conjunctions – often found in technological history – the early 19th-century technologies of the Jacquard loom and the electric telegraph merged with an obscure piece of mid-19th-century logic to become the basis of the mid-20th-century information revolution. Commercially, the early computer was essentially a massive upgrade of the tabulating–accounting machines of the 1930s; IBM, under Watson always adept at adapting technological developments to market possibilities, moved from the one into the other.

Conclusion: the pre-electric office

Thus the office of 1890 was already recognizable to modern eyes. Though it might still be gas-lit, though its workers might commute by horse tram, its secretaries took Pitman shorthand, transcribed it on typewriters, answered the telephone. The paradox remained that it might be entirely unconnected to electricity. And the office of 1935, by then electrically lit, might otherwise be remarkably unelectric. The connection remained to be made. After World War II, through the development of microelectronics, it was made with a vengeance. Meanwhile, early electronics developed largely in a separate, parallel line, albeit stemming, in part, from telephone technology.

Electronics: The scientific underpinnings

Technological historians agree that electronics only emerged as a distinct technology around 1900 (Tillman & Tucker 1978, Dummer 1983). The word 'electron' dates from about 1891 (Maclaurin 1949); the word 'electronics' only came into general use after the invention of the three-electrode valve in 1906 (Dummer 1983). But the scientific origins of electronics go back at least to the 18th- and 19th-century advances in physics, static electricity and magnetism, which have been summarized in Chapter 3.

More immediately, the birth of electronics resulted from scientific discoveries in the second half of the 19th-century, mainly in the UK and Germany. In 1864 John Clark Maxwell, a professor at the University of London, set out the mathematical rules governing electromagnetic phenomena, and predicted the existence of electromagnetic waves (Maclaurin 1949). Shortly after, Heinrich Hertz in Germany reproduced these waves experimentally; there were additional advances by Lodge in Liverpool and by Popoff in

Kronstadt. All these passed directly into Marconi's earliest experiments in wireless, around 1896.

In parallel, other scientists investigated the electrical properties of materials and the theory of the electron. At Marburg in 1874 Ferdinand Braun devised the detector rectifier, the first solid state electronic device, later to become the 'cat's whisker' of early radio; a little later, he developed the tuned electrical circuit (Braun & Macdonald 1982). In 1897 J.J. Thompson in England showed that cathode rays were particles of negative electricity, which he called 'electrons' (Handel 1967).

Other advances were to have a less direct, longer-term effect. The first was the discovery by Faraday, in 1833, that a few electrical conductors became better conductors as their temperature was increased, and that the opposite was the case. The second was made in 1839 by Becquerel, who noticed that a voltage occurred at the junction between a conductor and an electrolyte when the junction was illuminated. This is now known as the photovoltaic effect and it occurs in what are now called semiconductors. In 1873 Smith discovered that the conductivity of some substances changes under illumination, giving rise to what we now know as photoconductivity (Braun & Macdonald 1982).

These three fundamental discoveries contributed to the major theoretical advances in physics that occurred around the turn of the century and directly fed into advances in semiconductor technology in the succeeding decades (ibid.). The discovery by Roentgen of luminescence, whereby some solids emit light when illuminated by radiation of suitable energy, was followed up by Pohl and other German scientists in the years before and after World War I and led to important advances in semi-conductor technology. Indeed, Germany remained a major centre of fundamental work in semiconductor research between the wars (ibid.).

The working technologies

Without these scientific advances, crowded into the years 1864–1900, the technological breakthroughs could not have occurred. When Edison was puzzled by the problem of blackening in his carbon-filament electric lamps through use, in 1882, he noted that in certain conditions of vacuum and at certain voltages there was a strange bluish glow in the lamp, caused by an unexplained counter-current between the two wires in the lamp which formed the leads to the filament. The 'Edison effect' was the great inventor's only fundamental scientific discovery; but, finding no commercial significance in it, he did not pursue it (Handel 1967, *Electronics Magazine* 1981). It took intensive research by German and British scientists into the nature of cathode rays to show that the effect was due to the passage of electrons from the negative electrode to the positive. The theory of this thermionic emission process was worked out in 1903 by Richardson, one of Thomson's pupils, who showed that electrons are emitted from hot metals by a process similar to evaporation. These early discoveries and inventions 'led to the first two electronic devices which were to make radio and television possible, provide the basis for the electronics industry, and revolutionize technology – the thermionic valve and the cathode-ray tube' (Handel 1967, 54–5).

The thermionic valve, the first device in which free electrons were put to work, was invented in 1904 by Ambrose Fleming, 'the father of electronics' (Handel 1967). He was a professor of electrical engineering at University College in London, an adviser to the Edison Electric Light Company in London and subsequently a consultant to the British Marconi Company. He sought to develop a more efficient and reliable rectifier for radio waves than the existing coherer devices. His two-electrode valve, called a diode, was patented in 1904 on behalf of the Marconi company.

Fleming's pioneering device was somewhat crude; various inventors in different parts of the world soon registered patents for similar, improved devices. The decisive breakthrough came in 1906 with the invention of the triode valve by Lee De Forest, a radio manufacturer, in the USA. As the name suggests, this contained a third electrode or grid to influence the current which flows between the cathode and the anode; a small electrical signal applied to the grid could produce large changes in the current between the other two electrodes, thus providing the basis for an electronic amplifier 'which made possible the widespread use of radio and all that is associated with it' (Braun & Macdonald 1982, 13).

The triode was quickly followed by valves (vacuum tubes) with four, five or more grids; as the market and production grew, the tubes became smaller, more reliable and cheaper. Thus 'very shortly after the first experiments with wireless telegraphy, vacuum tube electronics had come onto the scene and determined the development of all electronics for the following fifty years' (ibid.). Electronics was thus very much a creation of the first decade of the third Kondratieff, 1897–1907; but, viewed in longer perspective, it formed a 'family' or subset within the wider family of electrical technologies; electronic circuits handle (progressively) smaller currents than traditional electrical ones and some incorporate 'active' components which are capable of modifying the flow (Soete & Dosi 1983).

The birth of radio

Valves were significant because – for over 40 years, until the discovery of the transistor – they were the basic components of radio. In a way typical of the cumulative process of invention, they appeared at precisely the right time; Marconi's first successful attempts at transmitting long-distance messages by wireless telegraphy came in England in 1901.

Radio, as one historian puts it, 'is not an entity, a thing in itself; it is simply the use of electromagnetic forces travelling in space. Invention in radio covers the designing of equipment for the better utilization of the properties of these forces' (Sturmey 1958, 15). Between 1830 and 1900 there were numerous attempts at communication by electrical signals without the use of wires (Tucker 1978); but they were necessarily short range systems relying on electrical conduction. Effective long-range wireless communication would depend on electromagnetic waves; and here, though Maxwell, Hertz and Braun had laid the scientific foundations, their potential for communication was largely overlooked by scientists.

As early as 1894 the young Marconi had begun to recognize and harness its potential for commercial wireless telegraphy, despite his limited scientific training (Tucker 1978). He began experimenting along the lines of Hertz's work at his home in Bologna; using an elevated aerial and an induction coil to generate sparks under the control of a telegraph key and an improved coherer for reception, he achieved a transmission range of 2.4 km. Finding little interest in Italy, and believing that England's large mercantile marine would prove the more profitable market for the discoveries he had made (Sturmey 1958), he came to London in 1896. With support from the Post Office, the Admiralty and War Office, he soon demonstrated radio telegraphy over several miles' distance. After forming the British Marconi Company in 1897 (at the age of only 23) to exploit his patents, his friendly relationship with the Post Office was terminated but he had sufficient commercial backing to continue his experiments. In May 1898 he successfully transmitted over a sea path of 23 km at Bournemouth and in March 1899 over a 50 km distance across the English Channel.

His goal, though, was transatlantic communication. Marconi, 'who was even better at "selling" wireless than inventing it', assured investors that most of the lucrative transatlantic cable business would be captured because of the much reduced capital and running costs of wireless (Handel 1967). In December 1901, Marconi attempted to transmit over a distance of 3500 km from Poldhu, Cornwall, to St John's, Newfoundland. Using an aerial hung from a kite to gain maximum height, Marconi and two assistants claimed to have heard a series of the letter S transmitted by prior arrangement from Poldhu, a claim then suspected but now generally admitted (Handel 1967, Tucker 1978). But his predictions of rapid commercial success had underestimated the difficulties of long-distance wireless communication; his company paid no dividends from 1897 to 1910.

Marconi then decided that the best prospects for early commercial success lay in ship-to-shore communications, where he soon won commercial success with contracts for equipping 32 British navy ships and a number of wireless stations around the English coast (Handel 1967). Soon the example of the British navy was followed by others; British Marconi, which insisted that its equipment should not be used for communications with other rival systems, was on the road to a world monopoly. This strategy was strongly resisted by German and American rivals and their governments; and in 1908 there was an agreement that international stations should be opened to all senders. But then, the sinking of the Titanic in 1912 dramatically underlined the value of wireless for safety at sea; soon after, legislation was passed in the UK, the USA and other maritime nations that ships above a certain size must carry wireless. This guaranteed further success for the Marconi companies (Maclaurin 1949, Handel 1949).

In this early field of spark wireless telegraphy, Marconi was thus the clear leader both technologically and commercially; by 1917 he had subsidiaries operating in about a dozen countries and had broadened the range of his communications business considerably (Maclaurin 1949). But this was a rather crude and limited method of communication, able to send the same kinds of messages through the air as had been sent for 80 years by telegraph wire; it could not compete with the telephone for speech transmission, where the real race now lay.

The idea of speech transmission by radio had occurred to several early practitioners in electrical science; in 1886 Leblanc in France suggested the idea of modulating a high-frequency current by speech signals (Tucker 1978). But the first practical demonstration came in 1902 from Reginald Fessenden, a Canadian who had been one of Edison's assistants and then professor of electrical engineering at Purdue University before joining the US Weather Bureau to develop apparatus for the wireless transmission of weather forecasts; here he made his first, unsatisfactory, experiments in 1900. Financed by two Pittsburgh businessmen to develop his work commercially, he concentrated on improving the high frequency alternator as a transmitter, and soon broadcast over a 25 mile range. After further improvements to the transmitter by Alexanderson of General Electric's research laboratory, Fessenden was able to broadcast over 100 miles in 1907. Alexanderson continued to work on the alternator over the next few years and raised it to a high degree of perfection (Jewkes *et al.* 1958, Handel 1967). Meanwhile, in Germany, research along similar lines was being carried out. AEG developed improved alternators in 1907 and Telefunken was active in improving the Poulsen arc (named after its Danish inventor) as an alternative method of transmitting speech (Jewkes *et al.* 1958).

The thermionic valve, and especially De Forest's triode, revolutionized radio. It not only improved the means of detection and reception but by 1911 it was being used for amplification of current and signals. This discovery was made independently by von Lieben in Germany and Armstrong in the USA in that year. The feedback circuit, to multiply the amplification by feeding the output of one tube into the input of the next, seems to have been discovered almost simultaneously in the USA, the UK and Germany in 1912. By 1913 the valve was being used to generate the high frequency oscillations necessary for radio telephony; soon after this, it superseded all other forms of transmitter. Again a number of workers have been credited, including Meissner of Telefunken in Germany, Armstrong, working independently, in the USA, and Round of British Marconi. Once the use of valves for transmission was understood, further progress in radio was largely a matter of improved valve design, although much effort for three decades after 1914 was also put into developing the specific field of short-wave radio. Thus in many respects the thermionic valve seems to have been the core technology in the development of speech transmission by radio, and hence the whole subsequent development of electronics technology. Its invention, and much of its early improvement, was independent of any commercial need (Jewkes *et al.* 1958; Sturmey 1958).

Here Marconi was strangely slow to see the opportunities: it was a team of Germans who initiated continuous-wave transmissions across the Atlantic in 1914; the following year the first wireless telephone signals were sent by the American Telephone and Telegraph Company (AT&T). Marconi's blind spot lay in not seeing the revolutionary possibilities of the Fleming diode, for which he had held patents since its invention in 1904; after the improved De

Forest triode of 1906, the application of electronics to wireless communication fell into American hands during a critical period, in which they established a head lead (Handel 1967, see also Jewkes *et al.* 1958).

World War I gave a great impetus to the use of valves and thus to their development (Sturmey 1958, Dummer 1983); by its end, triodes covered a wide power range and so were suitable for both receiving and transmitting purposes (Sturmey 1958). At the outbreak of the war, their commercial possibilities were being realized and research was being directed towards those commercial possibilities. During the war years, research was heavily directed towards military needs, especially in the UK, and ordinary commercial considerations were relatively unimportant. In the USA, however, the needs of war did not require such a complete cessation of non-military research and 'the effects of this were seen, for example, in the lead which America established in the field of radio-telephony' (ibid., 36).

Overall, it appears that in this early period, down to the outbreak of World War I, the leading companies in the radio industry were not American but British and German. The world market appears to have been dominated by Marconi and Telefunken, who between them controlled most of the ship and shore installations all over the world (Freeman *et al.* 1965). The various Marconi companies controlled a large share of coastal and marine installations throughout the world; the American Marconi subsidiary was the largest radio manufacturer in the USA down to World War I (Maclaurin 1949, Sturmey 1958).

It would be inaccurate to view these developments as the outcome of purely technological or market forces, however; governments seem to have recognized the strategic importance of radio communications networks quite early, and tended to support national firms and industries (Sturmey 1958, Freeman *et al.* 1965). This was seen not only in the British government's interest in developing an 'imperial chain' to link the various points of the then far-flung British empire but also in the involvement of the German and American governments. In Germany the electrical giants AEG and Siemens entered the radio communications field a few years after Marconi, but they quickly developed effective radio systems. With strong pressure from the German government, particularly the General Staff, they were persuaded to join their interests in 1903 to form the jointly owned Deutsche Betriebs-Gesellschaft für drahtlose Telegraphie, or Telefunken company. This firm was primarily concerned with R&D and with the sale, installation and maintenance of radio stations, but the parent companies continued to manufacture parts and equipment; by 1906 Telefunken had more installations than Marconi. Marconi's patents were not recognized in Germany until two years before the war, when Telefunken and Marconi reached a worldwide agreement on patents, licences and scientific knowledge (Freeman *et al.* 1965).

World War I 'changed the pace of development and "electronics" now covered a wider field of applications', as new radio tubes and new circuits were developed for communications, valves were mass-produced, new components such as resistors and capacitors began to emerge and assumed the basic form that they maintained down to the 1960s (Dummer 1983, 2). Although initial electronic applications and innovations had been taking place since the mid-1890s, it was only in the 1920s that electronics began to assume any industrial or economic significance.

Despite many pioneering inventions, the American radio industry was much less successful than its British and German counterparts until the formation of the Radio Corporation of America (RCA) in 1919. De Forest and Fessenden were not very successful as innovators or entrepreneurs, despite a string of important patents and orders from the American navy and war departments for equipment (Freeman *et al.* 1965). Large electrical firms such as General Electric and Westinghouse, and the telephone company AT&T, had major R&D resources and contributed to radio telephony. The early development of a strong indigenous American industry appears to have been blocked by the dominance of the British Marconi Company in international communications and by the opposing interests involved in the control of key patents. The way out of this deadlock consisted in the buying out of the American Marconi subsidiary and the setting up of a unified American-owned radio company (RCA). It was the result of a government initiative, impelled partly by an appreciation of the great commercial potential of radio and partly by the strategic imperative that a vital part of the communications network should not be controlled by a foreign, even if friendly, power. It was felt that in the postwar world radio (along with petroleum and shipping) was one of three key areas on which international influence would be based and that the USA should achieve a position in radio that was at least equal to that of the UK (Maclaurin 1949, Freeman *et al.* 1965).

In the precommercial period of wireless the bulk of the research advances were made by university scientists, and the conversion of these fundamental discoveries into practical commercial applications was made by young entrepreneurs such as Marconi and De Forest, who founded the earliest wireless companies in both the UK and the USA (Maclaurin 1949). Research was a haphazard undertaking, inadequately financed and dominated by the personalities of these inventors and pioneering entrepreneurs. But in the USA from about 1910 industrial radio research became of increasing interest to four large and well established electrical companies: AT&T, General Electric, Westinghouse and (from its foundation in 1919) RCA (ibid.). Here research was systematic, co-ordinated and well funded – proof, according to Maclaurin, of the Schumpeterian thesis that the monopolist (more accurately here, oligopolist) may play a creative rôle so long as he is faced by competition from potential new products and new technologies (Schumpeter 1942, Maclaurin 1949).

With the development after World War I of entertainment broadcasting – itself a key socio-organizational innovation – the inventor–entrepreneur became less significant; though a large number of firms were involved in radio manufacture, most of them operated under licences from the big firms; few carried out research, and both in the UK and the USA technical advance remained largely with a few major firms (Maclaurin 1949; Sturmey 1958). After Westinghouse joined the GE–AT&T–RCA agreement in 1921, RCA had no less than 2000 patents incorporating virtually all the important radio technology of the day (Maclaurin 1949). RCA bought 40% of its sets from Westinghouse and 60% from the General Electric Company (GEC). It also made some important international agreements, especially with British Marconi, Telefunken and the Compagnie Générale de Télégraphie, the dominant radio firms in their respective countries, which were to run from 1919 until 1 January 1945 (ibid.). Each company was to have exclusive rights to the use of the other company's patents within its respective territories as well as for what were called mutual traffic arrangements wherever possible throughout the world. Thus 'was organised the first international radio cartel' (ibid., 107).

Technical innovation after 1918 centred on the further development of valves and the use of short-wave bands, in which all the major British, American and German companies appear to have played a rôle (Jewkes *et al.* 1958). Another major innovation, frequency modulation (FM) to overcome static, came in 1933 from the independent American inventor Armstrong; major manufacturers were slow to react to this important advance (Maclaurin 1949, Jewkes *et al.* 1967).

The advent of broadcasting

More important in some respects were social and organizational innovations; in particular, the advent of broadcasting in the 1920s.

This came first in the USA, where there had been no wartime restrictions on amateur radio activity and where, consequently, radio sales to amateurs or 'hams' were increasing. In Pittsburgh, Westinghouse set up station KDKA for experimental broadcasts and these were so popular that a local store began selling large numbers of radio sets; accordingly, Westinghouse began the world's first regular broadcast transmissions in 1920, selling time to local advertisers; it started a boom in broadcasting and in the establishment of transmission stations, aided in its early stages by a complete lack of regulation (Sturmey 1958). This American lead was not based on technical accomplishment, yet it was a visible innovation which may have given the USA a lead in world markets for equipment.

In an attempt to counter it, Marconi in England secured a licence and constructed a station at Chelmsford in February 1920. It broadcast two daily half-hour programmes but then the licence was withdrawn. By this time broadcasting had also begun in several European countries but there appeared to be a strong official reluctance to initiate broadcasting in the UK (Sturmey 1958). Manufacturers, seeing the American developments, were eager; amateurs, and even some newspapers, joined them and the Wireless Society of London acted as a pressure group. Finally, in 1922 licences were granted for broadcasts in certain specified areas, mainly London and the other large cities; but they restricted output, apparently because of fears of interference with government stations and military manoeuvres.

No less than 24 companies wanted to set up stations, and a meeting was arranged at the Post Office in May 1922 to arrange co-operative agreements. Here two powerful rival groups of manufacturers appeared: one based on Marconi, GEC and British-Thompson-Houston (linked through the American patent-sharing arrangement); the other based on Metropolitan-Vickers, Western Electric (the British branch of the American Bell group's manufacturing arm), International Western Electric and RCC, an English company competing with Marconi in the marine radio field. Their main quarrel was over the Marconi group's claim for a monopoly of radio station construction. The resulting conflicts and negotiations delayed the start of broadcasting until November 1922. Finally, in January 1923 the British Broadcasting Company, guaranteed by the big six firms, came into existence. The Post Office agreed to issue receiving licences, one half of the proceeds going to the broadcasting company. When the company was wound up at the end of 1926 and broadcasting vested in a public corporation, the big six companies held 60 000 of the 71 536 issued shares and there were over 1700 other shareholders (ibid.).

At the start of regular broadcasting in Britain in late 1922, there were approximately 400 known manufacturers. Some were simply involved in reassembling surplus war radio equipment and many 'were pure dealers, often of doubtful honesty and even more doubtful technical ability ready to jump on any band-wagon which someone else started' (ibid., 143); there were also amateurs with considerable technical ability (including war-trained signals personnel) and established electrical manufacturers. The industry was very easy to enter, requiring relatively little capital, and it was estimated that between 10 and 20% of the manufacturers had no technical knowledge whatsoever; indeed, about half the radio sets in use in 1925 were of home construction. However, the industry grew rapidly; by 1925, four years after broadcasting began, it employed an estimated 40 000 people and had a turnover of between £10 million and £12.5 million (Sturmey 1958). The first electronics-based New IT industry had emerged.

Between then and 1939 fundamental research continued into the physics of semiconductor materials, such as the photovoltaic effect, and by the mid-1930s considerable progress had been made; in 1933 Pohl in Germany predicted that eventually vacuum tubes in radio would be replaced by small crystals, though it was almost two decades before that prediction was realized (Braun & Macdonald 1982). But additionally, major new practical advances occurred, which would have effect during and after World War II: television, sound recording and radar.

Television

Television represented not one but a series of technical inventions; the early basic concepts were developed by university scientists, but the later developments required large teams in well funded industrial laboratories (Maclaurin 1949). It developed along two very different lines: electromechanical and electronic. Early experiments with the former were carried out as early as 1884 by Paul Nipkow, and by a number of others including Professor Boris Rosing of St Petersburg, who succeeded in reproducing some faint images in 1907 and who included one Vladimir Zworykin among his students. By the 1920s John Logie Baird developed a mechanical system in the UK, and Ives of the Bell Laboratories produced one in the USA. The first public demonstration of television was given in the UK in 1925 by Baird using 30–60 scanning lines, to give a very crude picture by modern standards. The BBC began the first experimental TV service in the world in 1929, using Baird's mechanical scanning system. Broadcasts took place for half-hour periods on five days a week, but there was little public interest in these transmissions and sales of the Baird receivers remained small even though they came down in price from £40 in 1929 to £20 in 1931 (Handel 1967).

The key figure in electronic television was Zworykin. Emigrating to the USA in 1919, he joined the Westinghouse company, where in 1928 he developed the iconoscope, a device which transmits television images quickly and effectively. This removed a formidable obstacle to the development of commercial television. In 1930 Zworykin was transferred to the RCA Laboratories and by 1939 he and his team had developed a fairly sophisticated television system. Another major contributor to the development of electronic television was Philo Farnsworth, an individual and largely self-taught inventor who, with substantial financial backing, demonstrated an electronic system as early as 1927 (Jewkes *et al.* 1958). His largest single contribution was the image dissector. By 1935, backed by the Philco company, he had developed a much improved system.

Although many basic inventions in electronic television were made in the USA, others were made and first used for broadcasting in the UK. Electrical and Musical Industries Ltd (EMI) set up a research team which, working independently of the work of the RCA team in the USA, produced a version of the iconoscope, known as the Emiscope, and other improved equipment. By 1935, EMI and Marconi were able to offer the BBC a system on which regular broadcasting began in 1936, alternating with Baird mechanical system broadcasts. In the same year EMI exchanged patent licences and technical information with RCA. In 1937 a committee of experts in the UK voted in favour of the Marconi–EMI 405-line electronic system with Emitron cameras and cathode-ray tube receivers; regular programmes were broadcast by the BBC until the outbreak of war (Jewkes *et al.* 1958, Handel 1967).

Television required a large number of valves, and in order to reduce the size of television sets attention turned towards producing ever smaller and more reliable valves and other components (Sturmey 1958). It also called for the mass production of cathode-ray tubes. The first hard television tubes were produced commercially by EMI at the Marconi-Osram valve factory; the moulded tube was first used in the UK by GEC. All this needed substantial research money of the kind that could come only from major organizations like EMI or RCA. The part played by RCA is particularly notable (Freeman *et al.* 1965). Between 1930 and 1939, RCA spent \$2.7 million on television R&D, a further \$1.5 million on testing the system and about $2.0 million on patents and patenting; Telefunken and EMI probably spent similar amounts in this period (Freeman *et al.* 1965).

Radar

Radar, an abbreviation for radio detection and ranging, is a system of location which employs short-wave radio waves to detect objects. As with much else in electronics, its origins date back to earlier discoveries. Hertz's discovery in the 1880s that electromagnetic waves are reflected in a manner similar to light waves was important; in 1904 a German engineer, Hulsmeyer, was granted patents for a radio-echo device for preventing collisions; in 1922 Marconi and some American scientists independently sketched out plans to use the reflection principle for a device to prevent ships colliding with each other (Jewkes *et al.* 1958, Handel 1967). The first important application of the reflecting properties of radio waves was in measuring the height of various layers of the ionosphere, by Appleton in the UK in 1924 and by teams from the Carnegie Institute in the USA the following year. There was considerable research effort into the development of radar in the 1930s in France, Germany, the UK and the USA (Jewkes *et al.* 1967).

In the UK the government began to be interested in using radio waves to detect aircraft about 1934. Robert A. Watson-Watt, a senior researcher at the National Physical Laboratory (NPL), developed a working system; first tested on the Suffolk coast in 1935, it was installed along the east coast of Britain by the outbreak of the war, and had a decisive influence on the Battle of Britain and the subsequent course of the war (ibid.). Subsequently, the most important innovation was an airborne system allowing pilots to detect and intercept enemy planes. It required smaller aerials and thus shorter wavelengths and higher frequencies. The key invention here was the cavity magnetron, a thermionic tube of very unconventional design. First produced in 1939 by Randall and Boot, two physicists at Birmingham University, it too had a major impact on the war's outcome (Handel 1967). From 1942 to 1943 this allowed the development of sets that were more accurate and of greater range, permitting improved blind bombing techniques and naval search equipment (Jewkes *et al.* 1958, 1967).

Initially the American government and military were slower than their British counterparts. In 1940 a British mission went to the USA carrying the secret magnetron in a small box for further testing in a Bell laboratory, and the American US government set up the Radiation Laboratory at Masachusetts Institute of Technology (MIT) with a nucleus of 12 workers, a number that by 1945 had swelled to 4000. The magnetrons required the development of new magnetic alloys, new high voltage electronic switches, new types of conductors, improved cathode-ray tubes for display systems; in addition, non-active components, such as the transformers, attenuators, resistors and capacitors, had to be made to much more stringent and reliable standards for pulse operation (Handel 1967). All of this demanded a large team of researchers that only the USA could muster in these years.

Wartime work by radar researchers on the use of silicon and germanium as detectors directly fed into the invention of the transistor at Bell Laboratories in 1947 (Handel 1967, Braun & Macdonald 1982). Thus, although early work on radar was done more or less simultaneously in at least five countries, the UK had a convincing early lead in this area and in some related fields. The Germans, too, may have had a lead over the USA in semiconductor research down to 1940, though their radar R&D (mainly in Telefunken) remained one or two years behind the UK's throughout the war (Freeman *et al.* 1965, Braun & Macdonald 1982). But, with 30 or more American laboratories commissioned to research into semiconductors for radar purposes, by the early 1940s the UK's lead was unlikely to remain.

Thus work on radar, impelled by rearmament, had a major impact on the development of electronics technology generally. It led to a large increase in the numbers of researchers and engineers involved in electronics, which, even in the USA, had fallen in the depression and did not reach 1927 levels until 1942 (Maclaurin 1949, *Electronics Magazine* 1981). In Britain, at the peak of the effort, there were about 3000 people engaged in research and development work at government research establishments and another 1000 engaged in radar development work in industry. The high priority given to radar work, the direct co-operation of the best scientists from industry, universities and government, together with the relatively large resources available, 'led to an extraordinarily rapid flow of new devices and equipment, without previous parallel in the history of the industry' (Freeman *et al.* 1965, 55). Indeed, the success of the radar programme clearly shows 'how a technical lag could be overcome by a country with adequate scientific resources once the decision was taken' (ibid.), a lesson with possible relevance for the future.

7

The origins of British decline

The electrical revolution created a new set of industries – the high-technology industries of their age – in the UK, as in the USA or Germany; but from the start they grew more slowly than in these leading rivals, and the UK was clearly in third place. That fact was recognized by contemporaries, from the 1880s to the 1920s, who discussed the reasons for the British failure. In this chapter we first review the development of the British electrical and early electronics industries in the third Kondratieff, and then try to summarize the voluminous literature – both of the time, and by later historians – which sought an explanation.

British electrical manufacturing in a world context

Between the great innovations of the 1870s and World War I, electrical manufacturing in the UK grew considerably in production and employment. Census of Population figures (Table 7.1) suggest that employment in electrical manufacturing increased from about 2500 in 1881 to 49 500 in 1901 and to 94 000 in 1911. If we aggregate those involved in electrical supply and electrical manufacturing, we find a total of 108 000 'heavy-current' electrical employees in 1911; by including telephone and telegraph workers the total reaches 136 000.

These figures are based on occupations, not industrial sectors, and so are suspect. The first Census of Production, conducted in 1907, gives the first accurate official evidence on the growth and structure of the electrical industry in the UK. Between then and 1935 it appears that employment more than quadrupled whereas output, at current prices, rose nearly tenfold (Table 7.2). The late

Table 7.1 Employment in electrical apparatus and supply occupations, England and Wales, 1881–1911.

Occupation	1881	1891	1901	1911
electrical cable manufacture				5 813
electric lamp manufacture				5 627
other electrical apparatus makers and electrical fitters	2522	12 604	49 518	54 611
electricians (undefined)				29 970
electricity supply			2915	14 870
total electrical employment	2522	12 604	52 433	110 891
Post Office telegraphists and telephone operators	9422	14 955	22 819	10 338
non-government telegraph and telephone service				17 093
total electrical, telegraph and telephone	11 944	27 559	75 252	138 322

Source: Census of Population, England and Wales, 1911.
Note: These figures are derived from the occupational classifications of the Census of Population for the years concerned.

Table 7.2 Net output and employment in UK electrical engineering industries, 1907–35.

Year	Net output (current prices) (£m)	Employment
1907	6.8	62 300
1924	35.6	163 200
1930	51.6	213 600
1935	62.6	272 400

Source: Censuses of Production, 1907–35.

Table 7.3 Value of electrical manufacturing production, main countries, 1907, 1913, 1925.

Country	1913 Amount (£m)	1913 %	1925 Amount (£m)	1925 %
UK	22.5	13.9	65[a]	12.0
USA	71.5[b]	44.2	350	64.7
Germany	60	37.1	105	19.4
France	7.7	4.8	21	3.9
Total, 4 countries	161.7	100.0	541.0	100.0

Sources: Board of Trade COIT (1928a); Pope & Hoyle (1985); BEAMA (1929). Exchange rates (gold standard): £1 = $5 = 20 RM (Byatt 1979). BEAMA (1929) claimed a further £5.5 million production for the UK in the mid-1920s (BEAMA 1929).

Notes:
[a] 1924
[b] 1914.

1920s and early 1930s were years of fairly rapid growth according to some historians (Richardson 1967, Aldcroft 1968).

Despite this growth, electrical manufacture in the UK was slipping behind that of its major competitors. Comparative – indeed any – figures only become available from about 1890. As Clapham noticed long ago, the German census-takers did not distinguish a separate electrical industry in 1882; by 1892 they counted 26 000 workers in it: 'Beyond question, the creation of this industry was the greatest single industrial achievement of modern Germany' (Clapham 1936, 308). Even by 1913, however, Germany had conceded first place in production to the USA (Table 7.3), though

Table 7.4 Exports of electrical machinery and apparatus, five major countries, 1913–28.

Country	1913 £m	1913 % total	1924 £m	1924 % total	1928 £m	1928 % total
UK	7.6	23.4	16.2	30.4	18.3	27.4
USA	5.7	17.7	16.5	30.9	20.6	30.8
Germany	15.9	48.5	14.2	26.7	22.0	36.6
France	2.2	6.7	4.4	8.2	3.3	4.9
Switzerland	1.2	3.7	2.1	3.8	3.0	4.3
total	32.8	100.0	53.5	100.0	67.3	100.0

Sources: Board of Trade (1925); BEAMA (1929).

Table 7.5 Germany: employment in electrical and all industry, 1875–1965.

	Employment				
	All industry (000s)	Increase (%)	Electrical (000s)	Increase (%)	Electrical as % total industry
1875	4786		1		0
1882	5230	9	2	57	0
1895	6642	27	24	1237	0.4
1907	8177	23	117	380	1.4
1925	11 108	36	448	285	4.0
1933	7551	−32	252	−44	3.3
1950[a]	7637	1	454	81	6.0
1950[b]	4935		306		6.2
1955	6813	38	550	80	8.1
1960	8081	18	819	49	10.1
1965	8460	5	956	17	11.3

Source: Schulz-Hanssen (1970); see also Henning (1974).
Notes:
[a] German Empire.
[b] Federal Republic of Germany.

its exports still led the world (Table 7.4). Twelve years later, though production had increased in an absolute sense and as a percentage of total industrial production (Table 7.5), defeat and reparations had put German production relatively even further behind the American; and, though production continued to rise with the exception of the depression, and though during the 1920s its exports again overhauled America's, production would never again catch up. What is significant, however, is that the UK remained in third place. Though in production it narrowed the relative gap with Germany, as compared with the USA it fell farther behind; and its share of total exports showed a steady and ominous fall.

If it is difficult to distinguish early electrical employment or production, it is doubly difficult to distinguish different elements of the electrical industry. For Germany, we are on safe ground in saying that before 1880 virtually the whole industry was engaged on telegraphy and related areas. After that date the position changed drastically: Siemens & Halske, the leading Berlin firm, had some 88% of their turnover in the new activities related to electric power by 1896–7, and even by 1900, despite a remarkable two-and-a-half-fold increase in turnover in only four years, communications still represented only one-fifth of the activity in power generation and related activities (Czada 1969, von Peschke 1981). Still, in 1925 just over 26% of Berlin electrical employment was in the communications area; in 1939, after Nazi rearmament measures, nearly 37 per cent (Czada 1969). So it remained true that down to World War II, radio and communications constituted the main branch of the Berlin electrical industry (ibid.).

For the UK, Tables 7.2 and 7.6 summarize the evidence for the whole period from 1900 to the mid-1930s. These years saw continued expansion of the electrical industries. The war had clearly demonstrated the UK's technological and industrial lag behind Germany and other countries; the postwar reconstruction plans promised rationalization of electricity supply, promotion of electrical use, and the development of electrical manufacturing (COIT 1928a, BEAMA 1929, Corley 1966, Hannah 1979). There was some state support in the form of partial funding of an electrical industries co-operative research association and tariffs were imposed on imports of a few electrical products. Despite falling prices in the years following the 1921–2 crisis, the value of output of the main British electrical manufacturing industries reached about 190% of the 1913 level by 1927–8; consumption increased by 170% over the same period and exports by 150%; production for home consumption alone increased by 210% between 1913 and 1927, indicating improvement in the industry's competitiveness at a time of continuing difficulties for the German industry (BEAMA 1929).

Growth continued in the 1930s. The precise growth rates of the newer industries such as electricity in the 1930s are, however, a matter of major dispute between economists and historians. Thus Richardson asserts that electrical manufacturing output grew by about 82% and employment by 61% between 1929 and 1937, contributing to a significant economic revival in this period (Richardson 1967); employment in electrical manufacturing expanded much faster in the 1930s than in the 1920s, from 213 000 in 1929 to 259 000 in 1932 to 367 000 in 1937 according to this source (ibid.). Our data (Table 7.1), derived from the Census of Production but differing in detail, seem to be broadly in line, though his 1937 figure appears too high (see also Aldcroft 1970). The decline in world trade following the 1929 slump hit British exports, which fell from £19.4 million in 1929 to £10.0 million in 1933 and reached only £18.2 million in 1937 (ibid.); but protectionism after 1929 meant that imports fell from £8.4 million in 1929 to £2.6 million in 1933 and only rose to about £4.7 million in 1937, offering indigenous producers some protection from foreign competition (BEAMA 1929). Growth in the 1930s was also stimulated by the building boom, expansion of certain public utility schemes (especially railway electrification) and the stimulus of rearmament

Table 7.6 UK electrical undertakings, 1900–36.

	1900–1		1913–4		1925–6		1935–6	
	No.	Capital (£m)	No.	Capital (£m)	No.	Capital (£m)	No.	Capital (£m)
telegraph	24	31.2	31	35.3	32	42.4	26	37.2
telephone								
companies	16	8.9	20	14.9	11	8.2	14	12.8
government and municipal	3[a]	0.3[a]	4	13.3	3	62.3	3	114.7
supply								
companies	108	12.4	267	54.2	465	119.1	396	223.5
public	169	13.8	327	46.8	319	132.4	376	322.1
traction								
companies	75	26.1	186	168.0	126	186.7	66	207.9
public	18	2.7	173	47.7	151	84.6	85	72.7
manufacturing	135	21.7	284	47.1	376	91.0	346	107.1
miscellaneous	103	6.1	174	7.9	159	40.7	102	53.4
total	653	123.6	1466	435.1	1642	767.3	1414	1151.5

Source: *Garcke's manual of electrical undertakings*, 1938–9.
Notes:
[a] 1903
Data refer to operations in UK and Ireland plus a comparatively small number registered in the UK but operating overseas.

(Corley 1966, Richardson 1967, Aldcroft 1970). It was also due to the emergence of new electronics-based activities within the electrical industries, notably radio, but these constituted only a small portion of overall electrical manufacturing at this time. Thus, paradoxically, in the third long-wave downswing the UK seemed to have achieved rapid growth, catching up with the two leader countries both in the diffusion of electrical supply and power applications and in electrical manufacturing (see also Landes 1969, Aldcroft 1970). This was in part a result of the development of the national grid, in part of the rapid growth of the new electronics industries at the time – a phenomenon we now need to trace in the context of New IT generally.

The growth of new IT industries

Chapter 6 has shown that the last decades of the 19th century saw the emergence of a series of radically new information technologies, some dependent from the outset on electrical power, some to become progressively more so from the middle of the following century. The result is a technical problem for the researcher: the precursors of today's information technology industries were buried in a variety of industrial aggregates, the figures for which are not always detailed enough to separate out the relevant parts. So we cannot find relevant detailed data before the population census of 1921. What we can do is to treat the electrical industries as a kind of proxy for the embryonic new information technology industries, in the UK as in its leading industrial competitors.

By the first British Census of Production in 1907, electrical engineering accounted for about 1.4% of total net output of all manufacturing industries and about 1.2% of total manufacturing employment. Within this electrical sector, telegraph and telephone cables and accessories accounted for about 16.5% of total output (Census of Production: Historical Tables; *Garcke's manual* 1913–14). Electrical engineering employed 62 300. On the assumption that the shares of employment corresponded approximately to the shares of output of the various subsectors, then telegraph and telephone apparatus and cables employed about 10 500. In addition, Census of Population records indicate a total of 9440 in telegraph and telephone services occupations in England and Wales in 1881, a figure that had grown to 14 900 in 1891, 22 800 in 1901 and 27 400 by 1911 (Table 7.1). Most were engaged in running the information communications service, as well as in the installation, maintenance and repair of telegraph and telephone hardware, though a minority may have been in the production and manufacturing activities which are the major concern of this chapter. In 1907 the National Telephone Company employed 6049 wage earners and 979 salaried staff: a total of 7028 employees (Census of Production; *Garke's manual* 1913–14).

We cannot make similar estimates of employees in the production of new mechanical informational products, such as typewriters and calculators, at this time. But they are not likely to have been very high, since the UK was relatively slow to adopt such devices compared to the USA and other European countries; indeed, initial supply often came from overseas firms (Clapham 1938, Hays *et al.* 1969, Prais 1981). British inventors and innovators did not play a major role in the development of these devices (Margerison 1978).

The UK, however, shared in the diffusion of the new electronics technologies in the last decades of the third Kondratieff. By the 1935 Census of Production, total employment in these industries amounted to approximately 95 000, or approximately 1.7% of total manufacturing employment (see also Sturmey 1958; Wilson 1964).

For mechanical and electromechanical office machinery and document copying equipment, the 1930 Census of Production was the first to give figures. In 1930 about 5000 were employed, by 1935 about 13 000, and by 1948 about 22 000. Apart from duplicating machinery, British firms were not prominent among the initial innovators in these industries; the British market was still often supplied by imports or subsidiaries of overseas firms, although some domestic firms produced under technology licences from America and German firms. This reflects a general technological lag *vis-à-vis* other major industrial countries, especially the USA. This lag was evident not only in technological innovation and production but also, significantly, in consumption (Hays *et al.* 1969, Margerison 1978).

This innovatory lag was not replicated in the new electronics-based IT industries. As seen in Chapter 6, in radio the key initial innovator was Marconi, an Italian domiciled in the UK. Down to World War I, at least, Marconi's firm held many important patents and in many ways pioneered the development of wireless communications, although German firms and to a lesser extent American firms were also active. After the war, British firms such as Marconi were well placed technologically and organizationally to share in the growing market for both radio broadcasting transmitting and receiving apparatus and for international communications systems, naval radio systems and the subsequent application of electronic components in telephone systems (Sturmey 1958). Governments in the UK, Germany and the USA were keen to

Table 7.7 UK: major radio valve producers, 1919–39

Producers	Comments
Major companies in 1920s	
General Electric Co.	In 1919 combined with Marconi to form Marconi-Osram. In 1940s GEC took over MO Valve Co
British-Thomson-Houston Co.	Along with Metropolitan Vickers ceased valve making on formation of AEI in 1927
British Westinghouse (later Metropolitan Vickers)	
Marconi Co.	Joint owner of Osram. It sold its interest to EMI, which sold it to GEC in 1956
Edison Swan Electrical Co.	Later part of AEI (with BTH and Metropolitan Vickers)
A. C. Cossor	Ceased valve production in 1949
Mullard Radio Valve Co.	Founded after World War I with Admiralty backing. Taken over by Philips
Western Electric (later STC)	Mainly involved in valves for telephone systems
Major entrants in 1930s	
Thorn	
Ferranti	

Sources: Sturmey (1958), Allen (1970), Wilson (1964).

support indigenous firms in these fields and the UK's rôle at the centre of a vast empire, and trading and shipping activities served to enhance the growth of the electronics industry (ibid.).

In the radio broadcasting and receiving set industry, which grew rapidly after the establishment of broadcasting in the early 1920s, there were two major tiers or subsectors. Production of electronic components such as valves was relatively concentrated, because of the need for access to patented technologies and because of close linkages between electric lamp manufacturers, which characterized the early valve making industry and which continued (Table 7.7). But there were also new entrants, notably Mullard, set up by an ex-signals officer after the war with backing from the Admiralty, and Western Electric (later STC), which was primarily interested in valves for amplification of long distance telephone signals. In the 1930s new entrants to valve making included Thorn and Ferranti. In radio set manufacture, the industry structure in the 1920s was marked by hundreds of small firms and a few large electrical firms such as GEC and BTH. By the 1930s, however, the number of radio set making firms declined significantly; with the establishment of trade agreements within the industry, competition tended to shift from price to quality, and the firms that expanded most were those producing high-quality, sophisticated radio sets (Sturmey 1958, Wilson 1964, Allen 1970).

Between the wars, the British electronics-based IT industries were strongly placed in international terms. Their technological performance appears to have been more equal to that of the USA and Germany than in other electrical engineering sectors, where, as earlier seen, there was already a marked British lag. This favourable position in electronics was particularly evident in the 1930s, with the significant contributions of EMI and Marconi to the development of television and the early British lead in radar.

By this time, though, the American electronics industry enjoyed a long lead over that of the UK in productivity: output per person in the American radio industry was three or four times that of the UK, against a differential for manufacturing as a whole of two and a half times. This may have been partly due to the much larger American mass market; British performance compared more favourably in specialized equipment and components (Wilson 1964).

The electronics IT industries recorded very rapid rates of growth in the UK throughout the 1920s and 1930s (Table 7.8). Although precise official data are not available prior to the 1950s, fairly accurate estimates are possible using data from trade associations and early published studies (ibid.). In 1930 total employment in electronics (excluding telecommunications equipment manufacture and government departments) amounted to about 20 000; by 1939 this had trebled to reach 60 000. Gross output increased from £8.4 million to £27.6 million. The years immediately before and

Table 7.8 Radio equipment: exports, leading countries, 1923–7.

	UK	Germany (£000)	USA
1923	300	324	710
1924	465	960	1243
1925	1289	1478	2042
1926	1266	1581	1813
1927	1203	2099	1906

Source: Sturmey (1958).
Note: After the late 1920s, the growth of exports appears to have slowed due to the restrictions of trade following the economic crisis. In 1935 total UK exports of radio sets and transmitting etc. equipment, valves and other components amounted to £1.7 million (Wilson 1964).

during the war witnessed a very rapid growth due to military demand for radar and radio communications and navigational equipment, with output doubling between 1938 and 1945, and employment increasing from 54 000 to 91 000 over the same period (ibid.). Indeed, the war had a very positive effect in permanently raising the technical and production capability of the industry. Many of the firms which had been previously involved in television benefited greatly from wartime work on radar and the experience gained in the use of such devices such as cathode-ray tubes and time bases (ibid.). For many firms previously involved in the assembly of domestic receivers or gramophones, the experience of supplying the more diverse needs and demanding standards of the military led to much higher technical competence. Another effect of the war was that firms became increasingly accustomed to working with government establishments and departments as well as with each other; and this also led to the postwar industry becoming more highly organized into trade associations (ibid.).

So, to summarize: during the third Kondratieff the UK demonstrated an early and continuing competitive failure, relative to its main industrial competitors, in the electrical industries that were one of the main carriers of that long-wave. It showed an equal failure, albeit on a smaller scale, in the mechanical New IT products that also originated in the innovative wave of the 1870s and 1880s. The sole and shining exception was that at the very end of the third Kondratieff the UK demonstrated real strength in the crucial electronics products that were to provide a central basis for economic advance in the succeeding fourth-wave upswing. This notwithstanding, the question is why British industry should demonstrate such a critical innovative weakness in the electrical 'high-tech' field in this economic era, and in particular during the third Kondratieff upswing of 1897–1919.

The causes of British failure

Even before the third Kondratieff, but insistently during its first years, a perceptive minority began to warn British public opinion that the nation's industrial leadership was already under grave threat – above all in the new high-technology industries such as electrical manufacturing and chemicals. Lyon Playfair, a scientist MP, was one of the first to draw attention to the importance of scientific and technical education for economic development. In 1867, after studying European industry, he wrote: 'I am sorry to say that, with very few exceptions, a singular accordance of opinion prevailed that our country had shown little inventiveness and made but little progress in the peaceful arts of industry since 1862' (quoted in Court 1965, 168). He added that 'the one cause upon which there was most unanimity of conviction is that France, Prussia, Austria, Belgium and Switzerland possess good systems of industrial education for the masters and managers of factories and workshops, and that England possesses none' (quoted in Court 1965, 168).

At the start of the third Kondratieff, similar criticism came from the electrical inventor and manufacturer, Joseph Swann. In 1903, he emphasized that

> of the evil consequences of our educational deficiencies is the much less rapid progress that we, as a nation, have made, comparatively with our industrial rivals, more especially in those branches of industry which are the outcome of the scientific discoveries of recent times, and which largely depend for their evolution and successful practical application on original research and on the intelligent appreciation, by the capitalist and commercial class, of the resources of science and the advantages of high scientific training and scientific work as auxiliary forces in promoting industrial development and progress (ibid., 172–3).

The title of one book, *Made in Germany* (Williams 1896), neatly encapsulates the result. Germany's national product per capita then was growing at 21.6% per decade compared to the UK's 12.5%; her share of the world output of manufactured goods rose from 13% in 1870 to 16% in 1900, whereas that of the UK fell from 32% to 18% (Henderson 1975). In 1907 another critic, E.D. Howard, declared 'It is in the electrical industry that Germany has made her greatest progress, one of the direct results no doubt, of her excellent technical schools . . . This tremendous and rapid growth, which has placed Germany second only to the United States in the industry, was extraordinarily stimulating to other branches and accounts for not a little of the boom from 1896 to 1900' (Howard 1907, 58–62, quoted in Pollard & Holmes 1972).

From these critiques, and from subsequent analysis by historians, vital insights emerge: from the start of the third Kondratieff industrial growth increasingly depended on technological knowledge, an educated workforce, and managerial strata capable of harnessing the new knowledge. Additionally, technological change began critically to depend on the structure of industrial enterprise. Thus, at least from this point, technological history cannot be divorced from business history (Kindelberger 1978, Uselding 1980, cf.

Chandler 1962, 1975, 1984, Mowery 1983, 1984, Elbaum & Lazonick 1984, 1985). Henceforth, innovation depended no longer on individual genius or serendipity, but rather on systematic laboratory research within the business enterprise. It was here, rather than in some decline in the entrepreneurial spirit, that the beginnings of British competitive failure can be placed: British industry, in part because of the unique world hegemony it had acquired in the first and second Kondratieffs, failed to make the changes in business organization that were vital to success in the third. That was by no means determined; it was a failure of imagination and of the will.

Small scale of enterprise

Central in this was size of business enterprise. It has a two-way relationship to technological change. In the USA the transcontinental railroad and the telegraph helped produce a vast integrated continental market, facilitating the growth of the giant enterprise (Uselding 1980). But, in turn, such a firm could then necessarily develop a set of new managerial, administrative and technical functions, conducted by specialized and trained personnel. Critical among these was systematic industrial research which, by developing a stream of new products as well as improved old ones, would produce continuing technological leadership. Associated with it, specialized in-house commercial functions – distribution, marketing, advertising, after-sales servicing – provided critical feedback on the improvement of products. The resulting market power gave huge potential economies of scale, which, reinforced by process innovations that also emerged from in-house research, progressively lowered costs and further extended markets. And the advantages were most crucial in the leading-edge technologies, of which electrical manufacture was an outstanding example (Kindelberger 1978, Chandler 1984). This, notably, was the much misunderstood Schumpeterian view of innovation, though he was at pains to stress that large firms did not have a necessary monopoly of innovation (Chesnais 1982).

Let us develop this model. From the start, for its very existence the large firm will require early access to large financial resources, greater than those at the disposal of any individual entrepreneur. Only these can permit the establishment of mass-production facilities and of effective marketing and distribution networks. Such early access played a major part in the translation of major initial innovations (such as Edison's electric lighting) into rapidly growing firms on an international scale (Hughes 1983).

Secondly, the development and commercialization of the early major electrical product innovations often required a set of related technical solutions to radically new problems, constituting complete systems of innovations rather than single discrete products. This process required the efforts of a large research and technical staff able to tackle the various aspects of a complex set of individual innovations and deliver a comprehensive system that, though technologically radical, was yet safe and easy for the consumer.

Thirdly, the establishment and refinement of mass production and volume marketing techniques, rendered possible by access to financial resources and technical expertise, led not only to rapidly falling costs per unit of output but also to greater standardization of systems, products and components. This in turn aided rapid diffusion of the technology.

Fourthly, distribution and marketing expertise, together with after-sales service organizations, powerfully promoted demand for radically novel products, overcoming initial public suspicion and fear. This process, undertaken for commercial gain, nevertheless boosted the overall national market for products based on new technology.

Fifthly, within the larger firms a unique stimulus to product improvement and new product development came from the interaction between production-related research and development personnel and staff engaged in distribution, marketing and servicing functions. The latter provided feedback on users' experiences and satisfaction with existing products, thus aiding product development in line with market requirements, and facilitating rapid diffusion of new products.

The USA and Germany took the lead in this process in the first decades of the third Kondratieff. Although the large manufacturing enterprise with in-house research and engineering capacity and integrated marketing was initially an American innovation, it was quickly imitated by German electrical firms. As early as 1871, Siemens in Germany bought 'a shopful' of manufacturing equipment and set up an 'American shop' with mass-production methods; it also pressured customers to make standard orders, without alteration, thus permitting mass production of high-quality goods, in sharp contrast to the UK where the intermediary merchant encouraged idiosyncratic variations and inhibited long production runs (Kindelberger 1978). Between 1890 and 1914, helped by a financial crisis and bank intervention in 1901, 39 German electrical companies were effectively telescoped into two cartels under the leadership of the two leading firms, AEG and Siemens (Henderson 1975).

In the USA, similarly, the formation of General Electric in 1892 and the GE–Westinghouse patent agreement of 1896 effectively created three dominant electrical enterprises, GE, Westinghouse and Western Electric (Passer 1953). The impetus came from the growing complexity of electrical product innovations, impelled by the use of alternating current, and resulting in a shift towards increasingly complex systems of interconnected equipment (Passer 1953, Hughes 1983), coupled with process innovations requiring greater capital investment in machines and processes which were often themselves patented. In 1916 these three firms accounted for aggregate annual sales of more than $300 million, three-fifths of the estimated output of electrical goods of all kinds produced in the USA in that year (COIT 1928a, Passer 1953). In addition, these large American and German electrical enterprises had established a vast network of subsidiaries, associated companies, licensing arrangements and marketing units throughout Europe, South America and Asia including Japan (COIT 1928a, BEAMA 1927, 1929).

In comparison, British industrial organization remained small scale and fragmented (Hughes 1983, Mowery 1984), both in manufacture and in supply. Power supplies and electrical goods tended to be non-standardized and incompatible; even in the late 1920s, a multiplicity of local supply voltages necessitated a parallel multiplication of the kinds of electric lamp made, producing over 1000 varieties. At that time, belatedly, the Electrical Commissioners standardized future lighting circuits on 230 V (COIT 1928a). Thus the lack of adequate state regulation inhibited the achievement of economies of scale in transmission and generation, and in turn the suppliers suffered from the same failure. Research and development staffing and resources were meagre, and took the form mainly of collaborative industry-wide efforts which did not bring the same benefits as in-house effort (Mowery 1984). So economies of scale were few, product innovation weak, and market diffusion slow.

One effect was that already by 1912 two-thirds of the output of the electrical manufacturing industry in the UK came from subsidiaries of three overseas firms: Westinghouse and General Electric of the USA, Siemens of Germany. The rest was shared by a number of very small domestic firms; the other German giant, AEG, had a significant export trade to the UK (Byatt 1979, Chandler 1984). The German and American giants did not compete in each other's home markets, which were protected by tariffs (Byatt 1979); instead, they fought for a market share in third countries, especially the UK. This competition further eroded the capacity of British manufacturers to organize production on a standardized large-scale basis. It also inhibited product innovation and so distracted British firms that they failed to compete effectively in export markets (Garcke 1907).

Underlying this was an even deeper problem. The period from 1880 to 1914 was one of increasing internationalization of the economy, well marked in the new third Kondratieff industries and above all in the electrical ones. But this was internationalization of a new kind: it was no longer marked, as technological advance earlier in the century had been marked, by the free exchange of inventive ideas. The growing commercial potential of electrical technological advances, together with the growing consciousness of intellectual property rights, patent and licensing legislation, meant that international transfer of knowledge and technique took place in new commercial and organizational forms. The major inventions and innovations which took place in Germany and the USA were indeed quickly transferred to the UK, but largely through the commodified and commercial channels of either licensing and patent contracts, or imports or the establishment of production facilities by the leading German and American innovating firms. The UK, in other words, was becoming technologically dependent at the leading edge of development.

Some in the UK realized this. The Electrical Trades Committee of 1916–17 stressed the advantages of large-scale German production for her export trade, with branches all over the world (reported in COIT 1928a). Accordingly, the industry belatedly tried to reorganize itself: the 1920s and 1930s, the last two complete decades of the third Kondratieff, witnessed a wave of mergers in both electrical supply and manufacturing. By 1927 only 200 out of 508 registered electrical manufacturing and allied firms produced more than 80% of the electrical machinery and apparatus, other than cables; the four largest owned about 30% of aggregate capital in the industry, or, together with the 18 largest cable-makers, about half the total capital (ibid.). By then, the UK had three major groups of firms, with 58 000 workers and an output of £30 million; they had 60% of total output in heavy plant, about 35% of all production (BEAMA 1929). Specialist firms also merged into groups, thus strengthening their export capacity (ibid.).

Yet even then effective concentration appears to have been less than that in the USA or Germany (BEAMA 1927, 1929). The owners of large British firms continued to have a much greater influence on top management decisions; holding companies often remained little more than loose federations of family-run firms. 'Britain continued until World War II to be the bastion of family capitalism' (Chandler 1984, 498). Mergers, such as British Thomson-Houston and Metropolitan Vickers, in 1928, to form Associated

Electrical Industries, frequently resulted in a poorly co-ordinated collection of independent subsidiaries rather than an efficient, streamlined hierarchical structure, with rivalries between the units which had grievous effects for strategic management and co-ordination of research (cf. Jones and Marriott 1970, Mowery 1984). Thus, even as late as 1945, many manufacturing industries had failed to modernize in terms of size or production processes and mass production techniques (Hobsbawm 1969, Chandler 1984).

Neither the finance sector nor the state took a positive role in this merger process, contributing thus to its incompleteness and ineffectiveness. In particular, electricity supply remained organizationally very fragmented and geographically small scale until the 1920s, gravely hindering the achievement of economies of scale in the manufacture of equipment. The government restricted itself to promoting the formation of trade research associations, within which firms were encouraged to negotiate common pricing policies and production quotas. Paradoxically, this policy actually removed the motivation for mergers, preserved weak firms as well as strong ones, and reduced the incentives for British firms to pursue product and process innovations based on in-house research (Mowery 1984, Elbaum & Lazonick 1984).

Failure in research and development

By the end of the 19th century the major German and American electrical firms had established their own large research facilities. Systematic industrial research was indeed a late 19th-century innovation, which occurred in the electrical industry. Specifically, it occurred in Thomas Edison's Menlo Park laboratory, his 'invention factory' of 1876, where three years later he successfully demonstrated the incandescent electric lamp. Edison's 1093 patents also included several that proved crucial for the development of information technology, including the phonograph, the carbon telephone transmitter and the motion picture projector. The point was that this was specifically the world's first industrial research laboratory: it pursued research only for commercial gain.

Though Menlo Park closed in 1887, and though in the 1890s both General Electric (GE) and Westinghouse pursued innovation almost exclusively by purchase of patents, after the turn of the century GE and the Bell Telephone Company pioneered the large in-house research laboratory, devoted to fundamental as well as applied research (Passer 1953, Hughes 1983). Parallel to this, there was a development of fundamental scientific education and research in universities and polytechnics. There had been a massive expansion of higher education throughout the 19th century, but it was only towards the end of the century that the resulting expertise began to be applied and harnessed for industrial research. At the MIT the electrical engineering department opened with a four-year course in 1882; Cornell followed the year after.

In Germany the first chair of electrical engineering was established at the Technische Hochschule (TH) in Darmstadt in 1882, the same year as the foundation of the MIT department; also in the same year, the Berlin–Charlottenburg TH inaugurated a course in electrical machinery; the following year it established the first chair of machine construction and electrical engineering under the patronage of Werner von Siemens. For a number of years previously, von Siemens had been calling for the establishment of a chair of electrical engineering in all THs, for he felt that 'the need for young electricians grounded in science would exceed the number forthcoming from related fields or from informal education' (Hughes 1983, 144–5).

The UK, too, took an early initiative. The Central Institution of the City and Guilds of London Institute opened in 1884 offering courses in electrical engineering; it was headed by William Ayrton, a leading electrical engineering scientist and head of the physics department, who had toured German facilities before establishing the curriculum. Another initiative came from Sylvanus P. Thompson, the head of Finsbury Technical College and author of influential electrical engineering texts. John Hopkinson, a developer of engineering theory and a consultant to the Edison interests in England, was the first incumbent of the chair of electrical engineering at Kings College in 1890. The Central Institution of the City and Guilds, which changed its name to The Central Technical College in 1893, promoted and set examinations in electrical engineering for students studying at technical institutions throughout the country. As early as 1886, 151 candidates from various parts of the UK sat one of the two examinations in electrical engineering. By 1900 'the British student was receiving an education in electrical science and technology, that equalled the offerings at MIT and Cornell' (ibid., 155). The backwardness of the UK's electrical research, education and training was thus quantitative, not qualitative.

It partly lay in the lack of systematic industrial research. Here, the British story was very different from the German and the American. Early electrical innovation had received a powerful boost from the Royal Society and the Royal Institution of London founded in 1800 by an American scientist and adventurer

(Maclaurin 1949), which had made possible the career of Michael Faraday. But, throughout the 19th century, British scientific advance remained dangerously dependent on the brilliance of a few star performers rather than on the kind of organized research that became typical of Germany (ibid.).

The closing decades of the 19th century in the UK were marked by an awareness that it was falling behind Germany and other countries in scientific education and technological research, and that this would have disastrous consequences for military as well as industrial superiority. One notable response was the establishment of the Cavendish Laboratory at Cambridge University, in 1874, for the conduct of 'practical researches', over which James Clerk Maxwell (often called the scientific father of wireless) presided (ibid.). Towards the end of the century, the growing gap between the UK and some of the newly industrializing countries, especially in the new technology-based industries, gave reformers greater influence. To this end, in 1900 the National Physical Laboratory (NPL) was founded in direct imitation of its German predecessor in Charlottenburg (MacLeod & MacLeod 1977) with the intention of promoting the application of research to industry. The NPL carried out important research work for the British electrical industry in its early decades (COIT 1928a) and continued to conduct research on behalf of the industry's research association and the larger firms after they had established their own research laboratories (COIT 1927a).

But such efforts at restoring economic initiative through the support of scientific research were 'lacklustre in comparison with the example set by Germany' (MacLeod & MacLeod 1977, 304). Between 1900 and 1914 German industry and government had established several additional major industrial research institutions which dwarfed these British efforts of the time and enhanced further the relations between industry and scientific and technical research. In 1902 there were about 300 teaching posts for scientists in British universities in 1902, a further 250 places in government laboratories and perhaps 180–230 posts in British industry; in contrast, the German chemical industries alone employed more than 1500 scientists (MacLeod & MacLeod 1977).

World War I panicked the British government into greater intervention. One major result was the establishment in 1916 of the Department of Scientific and Industrial Research (DSIR). This assumed responsibility for the control of major civilian governmental research establishments such as the NPL, promoted fundamental research, sought to define national objectives in certain fields of applied research and worked to promote co-operative industrial research through research associations (MacLeod & MacLeod 1977, Mowery 1984). Between 1916 and 1920 the DSIR, acting through new research boards, encouraged the economic and industrial exploitation of knowledge in several sectors, including radio (MacLeod & MacLeod 1977).

A major new policy associated with the formation of the DSIR was the provision of financial support for industrial research associations (RAs). These were co-operative research organizations that firms within a given industry were free to join, to try to substitute for the relative lack of in-house research facilities in British industries. Over the first five years, 1918–23, 24 RAs were established. The expected industry support was not forthcoming and the RAs were government-funded until the budget cuts of the early 1930s forced major reductions. But in the rearmament years of the late 1930s and the subsequent war years, state support was significantly increased. Over their long history, however, the RAs had only 'a modest impact on innovation in British manufacturing' (Mowery 1984, 527). They not only failed to achieve independence from government aid, but they 'often failed to win active cooperation from industry' as firms were often reluctant to share valuable technical information and expertise (MacLeod & MacLeod, 1977 331–2).

Thus, though accurate figures are lacking (COIT 1927), available piecemeal evidence, together with patenting rates, suggest that Britain's industry was severely lacking in research compared to that of Germany and the USA. Indeed, even as late as the 1950s, when comprehensive data were available, comparative studies show that industrial R&D lagged severely behind American industry (Freeman 1962). In the mid-1920s the Committee on Industry and Trade estimated that the amount spent annually on all industrial research in the USA amounted to $75 million and the amount spent by American manufacturers in laboratory research alone amounted to $35 million (COIT 1927a). It reported that GE in the USA spent $3 million on in-house research annually, and that Bell Laboratories, together with the research department of the American Telephone and Telegraph Company, employed a staff of over 4000 persons, of whom some 1800 were trained engineers, chemists, physicists or other scientific workers engaged exclusively on R&D type activities (COIT 1927a). Another source suggests that by 1926 industries in North America were spending about $200 million on applied research and development and employing over 30 000 scientists (MacLeod & MacLeod 1977). The Committee on Trade and Industry underlined 'the astonishing progress made in Germany in industries where technical knowledge and scientific methods are essential' as clear evidence of the relatively high levels of research taking place in companies, universities and other organizations in that country (COIT 1927a 320). It confirmed the insignificance of the research of a typical British trade association, with an income of perhaps £5000 to £6000 a year, in comparison with the giant German and American industrial laboratories (COIT 1927a). In fact, much of the work of industrial scientists consisted in routine testing rather than long-term strategical research (COIT 1929).

The Committee was unable to provide comparative data on the levels of in-house electrical research in the UK. It did, however, estimate that member firms of the British Electrical and Allied Manufacturers Association (BEAMA), who produced about 80% of the national output of electrical manufacturing, spent about £250 000 in aggregate on research annually; of this, the British Electrical and Allied Industries Research Association had a total budget of £100 000 over the five years 1920–5. Although these figures do not include work by the government research establishments related to the electrical industry, or in-house research by the three largest firms, they are tiny compared to the sums indicated for the General Electric and Bell Laboratories in the USA (COIT 1927a, 1928a). Thus, the Committee concluded: 'the expenditure on research by the British industry is relatively small in comparison with the similar expenditure of American firms' (COIT 1928a, 291), especially on broader and more fundamental research.

However, the slump following 1929 meant a loss rather than an expansion of research activity. The DSIR and universities' budgets were cut, and research programmes suffered from stop–go policies. One 1934 survey revealed an unemployment rate of 10% among graduates in 'technical work'. Only in the latter half of the 1930s did rearmament again stimulate R&D expenditures. By the end of the decade the number of industrial scientists had increased to 4382 in about 520 firms (MacLeod & MacLeod 1977).

Yet there was a parallel growth in Germany and the USA, so that the relative gap between the countries remained very large. One 1939 estimate, by J.D. Bernal, showed R&D expenditures by British industry of £2.2 million (which, because of lack of data, may have been an overestimate) and total research expenditures by the British government (including its support of the research associations) of £2.95 million (Bernal 1939); most of the industrial work took place in the laboratories of a very few firms. In contrast expenditure on research in American industry was $175 million in 1937; total federal research expenditure was $72 million, of which half was spent within industry (Mowery 1984). In 1933, 116 in a sample of 160 of the top 200 American firms had research laboratories, but only 20 of the largest 200 British firms had such facilities in 1936. Total American research (academic, government, industrial) at the end of the 1930s was perhaps $300 million or ten times the British level (Bernal 1939).

Bernal in 1939 already underlined the consequences: 'Few results of any fundamental importance have in the past emerged from any British industrial laboratory, whereas many such have come from German and American laboratories; [there was] an undeniable tendency to import scientific results ready-made rather than to develop them in English laboratories' (ibid., 56–7). The reasons lay deep in national culture: the British scientific tradition, pragmatic and antitheoretical, had been of use in earlier simpler ages but was rapidly becoming unsuitable to the challenges of the 20th century (Bernal 1939).

The rôle of finance

The rapid industrial development of Germany and USA in the third Kondratieff took place not only behind a tariff wall – on the advantages and disadvantages of which economists and historians still disagree – but also with the active co-operation of the banking system in mobilizing resources and channelling them to the productive enterprises, thus powerfully aiding the development of the dynamic, large-scale, integrated corporate enterprise (BEAMA 1927, Landes 1969). In the UK, finance capital became increasingly overseas oriented; in Germany, it remained at the service of industry (BEAMA 1927). This was particularly important for the development of the electrical industry. In the USA, ready access to large financial resources, together with active participation of the financiers themselves, greatly facilitated the creation of demand for the products of the new industry – for instance, by the supply of credit to would-be electrical utility companies (BEAMA 1927, 1929, Passer 1953, Chandler 1975, Hughes 1983). This extended into overseas markets, where GE and AEG were particularly active in creating supply companies in Italy, China, Japan and South America (BEAMA 1927). Despite the overseas involvement of British finance capital, it generally did not benefit British manufacturers to the same degree (BEAMA 1929).

Conclusions

For the relative British competitive failure in the electrical industries during the third Kondratieff, therefore, the central explanation lies in the inability to develop strong, large-scale corporately managed enterprises with integrated R&D and marketing functions, such as emerged in both Germany and the USA. Though Britain's adherence to free trade whilst competitor industries in other countries were protected may have provided part of the reason for this, other factors were probably more important: the lack of large, centralized, standard electricity supply systems (associated, in part, with the existence of a relatively efficient and cheap gas supply system); the weak and passive support from the financial sector; and the failure to develop mass technical education. Here, as so often in economic history, the follower countries seemed to possess some inbuilt set of advantages over the old-established leader.

The central question, important in its implications for subsequent history, concerns the rôle of the state. In Germany it was all-pervasive; in the USA, far less so. The state in Germany played not merely a direct rôle (in military procurement, in technical education) but also a more indirect and subtle one, acting as orchestrator and co-ordinator of different sectors of capital; obvious comparisons are with French economic planning in the 1950s and 1960s, or with Japanese industrial policy after 1950. In the USA a minimalist federal government did relatively little to influence the growth of the new high-technology sectors save, perhaps, during and immediately after World War I and the run up to World War II; state governments provided generous systems of technical higher education, but that was about all. Yet both prescriptions proved outstandingly successful. In particular, both avoided the problem that plagued and greatly retarded the development of electrical manufacture in the UK: the fragmented small-scale nature of the supply industry and the consequent failure to standardize.

Finally, we should underline the point that during the 1930s the British electronics industry appeared the outstanding exception to this gloomy conclusion. Organized partly into a few very large-scale companies with major international markets, partly into a series of small but innovative components firms, it seemed to be all that the older electrical industry was not: it was growing, it was technologically advanced, it seemed to be competitive internationally. It promised to take the UK into the technological forefront of the next long-wave upswing. Its post-World War II history will form the theme of Chapter 12.

8

Electropolis: Berlin, Boston, New York and London, 1847–1952

In the second and even more so in the third Kondratieff, innovation came out of the great metropolitan cities. Above all it came out of Berlin, the city that from 1880 to 1914 was the high-technology industrial centre of the world. There are curious parallels between the Berlin economy of that era and the high-technology localities that in time supplanted it: Massachusetts' Highway 128, California's Silicon Valley. In both the earlier and the later examples there are striking similarities in the rôle of the state and public utilities, above all of the military machine and telecommunications agency; the critical importance of organized research both within and without the firm; the needs of industry for skilled labour; and the rôle of technological innovation in national economic advance.

The Boston – New York axis after 1870 became Berlin's main rival on the electrical world stage, not least because of the innovations of Thomas Alva Edison. For a variety of reasons, both its basic research and development and its production facilities were soon decentralized to rural and small-town locations. But it demonstrates the extraordinarily close relationships between the information needs of a great commercial market and the early history of New IT, and how this new technology in turn produced new innovations, not themselves of direct importance to IT but indirectly of incalculable future significance.

London in that period was a pale shadow of Berlin. It, too, was an electrical metropolis, dominating the British industry. But its smaller rôle reflected the declining world importance of that industry and in turn its weakening potential for innovation, which we have followed in detail in Chapter 7. Its value is chiefly comparative: it shows how, during the period down to World War

II, the great metropolitan city was still the home of the innovatory high-technology industry of that day; it thus poses the question of how that rôle has since been lost.

The story is complicated by one fact. As already emphasized in Chapters 5 and 6, information technology was only one part, and for a time a decreasing part, of a whole complex of innovative electrical and other engineering industries, in which – between 1880 and 1914 – Germany competed with the USA for world technological and economic leadership. Germany's greatest technical advances came outside the information technology area, though it made many important contributions in that area also. But the important point is that electrical technology proved to be a seamless web, in which advances at one corner proved to be of crucial later significance in another. So it is necessary to tell the whole story, even though parts of it may at first sight seem irrelevant to the main theme.

The key to it is that, as seen from Chapter 5, there are two main kinds of electricity, which the Germans usefully distinguish as *Schwachstrom*, weak current (as used in communications devices like the telegraph and the telephone) and *Starkstrom*, true power current. The first technical advances came in weak current for telegraphy and related devices, in the 1830s and 1840s. Further major jumps came in the 1870s with the telephone and in the 1900s with wireless telegraphy, leading on to the development of modern electronic information technology. But between 1880 and 1910 the most sensational carrier technology advances were made in the development of electric power for lighting, industrial power and transportation.

Berlin: Elektropolis

Although Berlin was among the leaders in early weak current and continued to innovate in it, its greatest triumphs came at this time in the development of power. In the 1890s the electrical industry became 'an enormous motor which in those years drove the German economy forwards' (Siemens 1957a, 193), helping powerfully to produce technical progress that, in the estimates of later economists, contributed between 30 and 40% of the total growth of industrial production (Andre 1971, Schremmer 1973). It was thus, between 1880 and 1913, that Berlin became the electrical capital of the world, Elektropolis (von Weiher 1974). By that latter date it was the leading industrial and commercial city of Europe, regarded by

Table 8.1 Leading centres of the German electrical industry, 1925–39.

	Employment (000s)		
	1925	1933	1939
Berlin	174.6	83.4	235.2
Stuttgart	18.9	12.5	28.3
Nürnberg	15.2	10.1	24.3

Source: Czada (1969).

contemporaries as the ultimate example of the progressive city (Hughes 1983).

Whatever the branch of the German electrical industry, whatever its importance or unimportance, Berlin dominated it. Berlin had some 52% of total German electrical employment in 1895 and again in 1907, 50% in 1925, 42% at the end of the Great Depression in 1933, 50% in 1936, and 44% in 1939 (Czada 1969). Throughout this period Berlin was the greatest single centre of electrical production in the world (quoted in Fischer 1976). In 1925, 37% of employment in telegraphs and telephones, over 63% of that in radio, and nearly 45% of employment in communications devices was found in the capital (Zimm 1959). Fourteen years later, just before the outbreak of World War II, close on 56% of employment in telegraphy, radio and telecommunications was still in Berlin; this was a far higher proportion than any other sector (Czada 1969). In comparison to Berlin the other leading centres of the industry were puny (Table 8.1).

Inside Berlin, the industry was born as a highly monopolistic one because of its huge capital needs (Holthaus 1980). Two giant organizations dominated: the older, founded in the 1840s, the family firm of Siemens & Halske (S&H); the younger, created in the 1880s, the Allgemeine Elektrizitäts Gesellschaft or AEG. Despite its

Table 8.2 Employment in S&H and AEG, 1880–95.

	S&H	AEG
1880	800	—
1885	1100	—
1890	3350	2000
1895	4070	5100

Source: Czada (1969).

early start, the older firm was rapidly overhauled in the early 1890s (Table 8.2).

Two revolutions: Berlin's electrical industry in the second and third Kondratieffs

The first technical advance, which established the firm of S&H and with it the Berlin electrical industry, was the invention of the electric telegraph by Samuel Morse in the USA and by Cooke and Wheatstone in the UK, in 1837; Werner Siemens developed and refined their work from 1846 onwards, going into partnership the following year with a technician from the Physical Society of Berlin, Johnann Halske. We saw in Chapter 4 that their first workshop, in the Kreuzberg area close to the Anhalt railway station, was significantly close to the headquarters of the War Ministry in the nearby Wilhelmstrasse, from which their early telegraph activity drew much of its business (Siemens 1957a, Barr 1966). The location was somewhat similar to the Pimlico area of London, where also small engineering workshops depended on government orders from nearby Whitehall ministries.

It was two decades later, in 1866, that Siemens, among others, invented a workable dynamo. He immediately grasped its practical significance; in a paper read to the Physical Society he said that it would make cheap and convenient electrical power available to industry everywhere that labour was available (von Weiher & Goetzeler 1981) and, in a letter of 1866, he even identified it as the pivot of a great technical revolution (Siemens 1957a). In 1878 Siemens developed a practicable electric arc lamp, in 1879 a trial electric railway, in 1881 the first electrical tram service. It was from this time that the real expansion of the firm began: between 1881 and 1890 it grew from 870 to 2909 employees (Kocka 1969).

As it occurred, the two sides of the business developed in parallel. The old works (the 'Berlin works' of 1851 on the Markgrafenstrasse in Kreuzberg, close to the original establishment) continued to make telegraph equipment, whereas a new (1883) works at Charlottenburg, close to Werner Siemens' villa (to which he had moved in 1863), made power current equipment (Siemens 1957a).

Now, however, a rival appeared on the scene. Emil Rathenau, born of a Berlin business family in 1838, trained as an engineer.

After the failure of his first business, he grasped the significance of electrical lighting and was quick to secure the German rights on the incandescent lamp, which had been invented by Thomas Edison in the USA in 1879 (Anon. 1956). His original interest was in electrical installation and electrical supply; but he found that he needed the expertise of Siemens to build the necessary equipment. In an agreement of 1887, the newly formed Allgemeine Elektrizitäts Gesellschaft retained manufacturing powers but undertook to buy larger equipment from S&H, which in turn had to offer major lighting contracts to AEG (Siemens 1957a). Thus, almost from the beginning of the new giant, cartel formation was the rule.

From then on AEG concentrated especially on the growing heavy side of electrical engineering, which mushroomed in the 1890s through the rapid development of electric lighting and electric traction (ibid.). The key to this second electrical revolution was the conversion of direct to alternating current, allowing the generation of electric power in large central power stations and its transmission over long distances, as first successfully demonstrated at the Frankfurt Electrotechnical Exhibition of 1891 (Schulz-Hanssen 1970). It was AEG that made the critical innovations: the AC motor and generator, and the transformer, during 1889–90. So it made the running in power supply. The major polyphase power stations, at Moabit and Oberspree, were built next door to the AEG works, the latter by AEG itself, which owned the Berlin power company (BEW) until it sold it to the city in 1915 (Hughes 1983).

Siemens took second place here, but it continued to lead in information technology. In the 1880s Siemens prospered hugely from the introduction of electric blocks and electric motors for railway signals and signal boxes, with a growth in output in this division from 350 000 RM a year between 1880 and 1890 to 1.7 million RM in 1896 and 8.2 million RM in 1907 – a nearly thirtyfold increase in only 20 years (Czada 1969).

In the production of telephone and telegraph equipment, too, Siemens kept a commanding lead, with over 80% in 1933 (ibid.). In the 1890s it benefited from the early growth of telephones (Siemens 1957a); in 1907, despite its relative lack of R&D capacity, the Post Office entrusted it with the work of conversion to automatic exchanges (ibid.). True, the German Post Office took care to keep a number of smaller competitive firms – Lorenz, DeTeWe (Deutsche Telephonwerke und Kabelindustrie AG), Mix & Genest – in business, but in certain areas, such as the development of self-dial automatic exchanges, S&H had an effective monopoly until the Post Office welcomed a patent association of other producers in 1922 (ibid.). Still, from 1908 to 1928, Siemens's sales of exchanges served some 1.5 million subscribers, one quarter of the world's total production (Siemens 1957b).

Siemens also continued to innovate in telegraphy – spurred on by the demands of the German navy, who gave it a virtual monopoly of production (ibid.). It developed a high-speed telegraph which used alternating current; it developed the same system for underwater telegraphy, thus increasing capacity on overloaded international lines such as that between Germany and Sweden; it produced a device for transmitting newspaper pictures. And it developed teleprinter (telex) machines that could be operated in ordinary offices, though a conservative Post Office administration resisted their introduction until 1933; ironically, the system then provided a vital information network for the Nazi state (ibid.).

S&H was, however, slow in moving into radio. In 1903, encouraged by the Kaiser – who was concerned about the military importance of the new technology – it made a 25-year agreement with its rival AEG to develop radio jointly through a new subsidiary, the Gesellschaft für drahtlose Telegraphie mbH, which soon became known by its less cumbersome telegraphic name of Telefunken (Siemens 1957a). At first the technology was slow to develop. But World War I gave an enormous boost to the radio industry by installing thousands of sets and training hundreds of thousands of men in their use. During the 1920s important technical advances – the loudspeaker and the amplifier – created the modern domestic radio. The USA, with its commercial freedom, exploited the new technology more effectively than Germany, where the Post Office had a regulatory monopoly. Nevertheless, there were 3 million licensed radios by 1928 – a figure second only to the USA. (Siemens 1957b); by 1933 nearly 78% of households had radios (Czada 1969). Both S&H and AEG depended for manufacture on Lieben electron tubes, developed and patented by Robert von Lieben in Vienna in 1906–10, on which, until 1933, Telefunken had a monopoly; thus, both firms found themselves in the extraordinary position of being licensees of their own subsidiary (Siemens 1957b).

Radio, essentially, was a new field for Siemens: the firm's whole previous experience in telegraphy had been in supplying capital equipment to large government monopolies, but now it had to enter the consumer market. It was perhaps appropriate that Siemens chose to do it not directly, but through a subsidiary. In fact, during the 1920s two competing divisions used the same equipment in different cabinets (ibid.). It was this new stress on mass production that provided the main drive to move out in search of more space (Siemens 1975a).

Siemens was less slow in moving into other new information technologies. In the 1920s it worked together with AEG to develop the Klangfilm system of talking pictures, which then fought a bitter legal fight with the rival Tonbild-Syndikat until the two agreed to exchange patent rights. In 1930 followed a world agreement, the celebrated 'Film Peace of Paris', with rival American patent holders. Interestingly, a group of inventors outside the Siemens–AEG orbit, the so-called Triergon group, had developed and demonstrated a successful sound film technique in Berlin as early as 1923, but had been frustrated by the artistic conservatism of the German film industry (Siemens 1957b).

The location of the industry within Berlin

Within Berlin the electrical industry followed a classic course (Fig. 8.1). It started in small home workshops, close to the centre of the city. As it grew, especially in the explosive years after 1880, it moved out to the then edge of the city in search of space. But in this process it always found itself on the horns of a dilemma: the further it moved out in the search of plenty of cheap land, the more difficult it proved to attract labour.

Siemens and Halske started their partnership in 1847 in a combined house and workshop in the Schöneberger Strasse in the inner-city area of Kreuzberg. There seems little doubt this location, as in the case of the second similar establishment to which they moved in the nearby Markgrafenstrasse in 1852, was chosen for its nearness to the contracting offices of the Prussian Post Office and the War Ministry a few blocks away. The first major move of S&H – to the new works at Salzufer in Charlottenburg, at the western edge of the city, in 1883–4 – probably had three major motives. Firstly, Werner Siemens had built his villa in Charlottenburg some 20 years earlier; secondly, the associated firm of Gebruder Siemens had set up here in 1872; and thirdly, the location was close to the research facilities of the Technische Hochschule, the precursor of today's Technische Universität, which is still on the same site.

At this time the older information technology part of the firm's business was still in the 'Berlin Works' on the old site, but the newer activities (cables, dynamos, lamps) went to Charlottenburg (Zimm 1959). Thence, it was logical for the firm to expand outwards from the Charlottenburg site as transport facilities became available. In 1899 the first plant, a cable works, was established on the Nonnendamm in Spandau, then an area of open land some

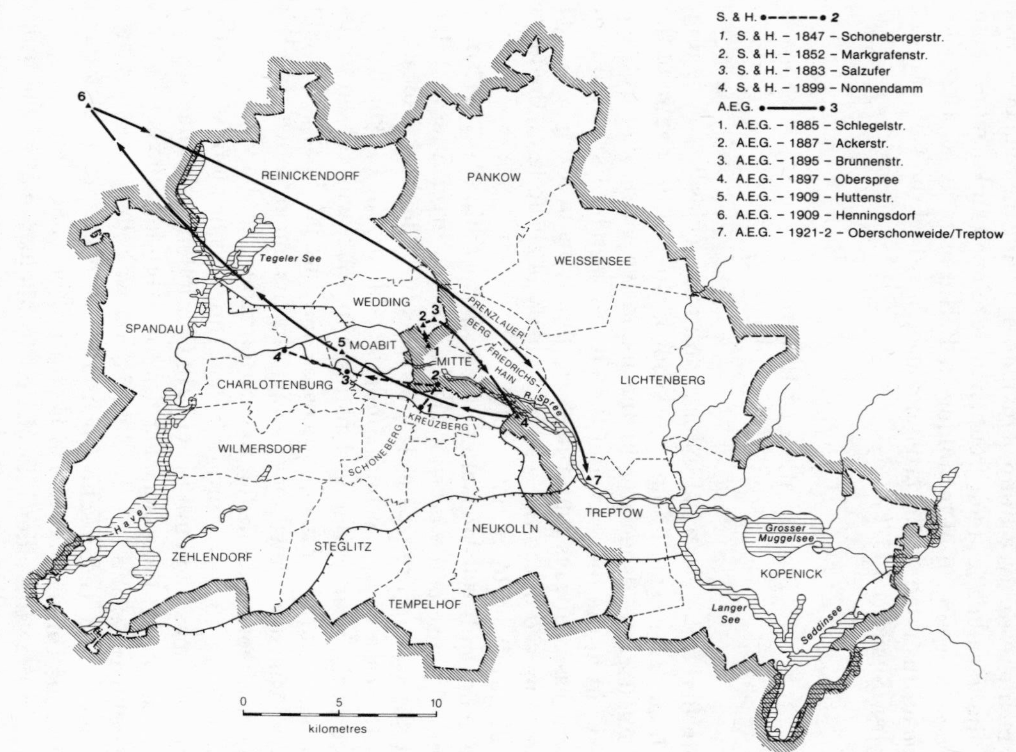

Fig. 8.1 Greater Berlin: electrical industry, key locations, 1842–1921

three-quarters of an hour's walk to the nearest habitation (Siemens 1957a). From then, the site was expanded almost continuously until 1912, when the headquarters offices were moved; telecommunications had already moved to the new Wernerwerk there by 1905 (ibid.); by 1914 there were nearly 21 000 factory workers on the site with another 2000 office workers (ibid.); by 1929 there were over 57 000 (Zimm 1959). But by this time the firm had embarked on the development of a vast housing project next to the works, employing the Weimar republic's leading modern architects to design it.

AEG followed a similar course, albeit at greater speed, reflecting this firm's much later start. Its first establishment of 1885 was in the inner city at Schlegerstrasse close to the Stettin (now North) Station. Within two years it had moved out to the working-class inner suburb of Wedding; in 1895 it established another plant close by. A critical factor was that Wedding was in the centre of a large area of working-class housing; AEG's director, A. Riedler, said in 1916 that it was possible to recruit 100 skilled workers a day there (Czada 1969). At this time it looked as if AEG was following a radial line of movement northward out of the city, similar to Siemens's progress. But subsequently AEG went in different directions: it developed a big complex in the south-east of the city, along the banks of the Spree in Nieder- and Oberschönewiede, while developing another complex at Henningsdorf outside the city boundaries to the north (Zimm 1959, Czada 1969). AEG in fact followed a strategy different from that of Siemens; whereas Siemens moved out and abandoned its old works, AEG simply multiplied plants at different locations, the new factories taking the fast-growing activities from the older ones (Czada 1969).

Generally, then, down to the early 1880s the industry was still heavily concentrated in the inner city. A first period of migration, from then until 1895, caused it to cluster at about 5 km from the centre – the circle formed by the ring railway. Here were found the Siemens plant at Charlottenburg, the AEG plants at Wedding and at Treptow (Zimm 1959). Then, between the mid-90s and World War I, a second phase of expansion and migration, aided by the extension of the city's own passenger transport system, took both major concerns further afield, in search of cheaper land and good transport links (ibid.). There were undoubted risks in this strategy at first; before the transport links developed, the vital skilled workers might not follow. For this reason, neither of the two giants could afford to move too far (Czada 1969). So the outward displacement did not take them (or their smaller competitors) right out of Berlin. Rather, the industry moved progressively, in stages,

Table 8.3 Location of the Berlin electrical industry, 1895–1939.

	Percentage of total Berlin employment				
	1895	1907	1925	1933	1939
Old Berlin	68.5	61.8	39.8	31.6	31.2
Charlottenburg	27.6	14.2	7.8	7.8	5.6
newly incorporated areas	3.7	23.5	50.6	55.8	59.3
(Spandau)	(—)	(15.5)	(27.3)	(26.7)	(22.9)
outer area	0.2	0.5	1.8	4.8	3.9

Source: Czada (1969).

within the city (Table 8.3). It did so predominantly along the banks of the River Spree, north-west and south-east of the inner city, where land was cheaper; thus, this early precursor of Silicon Valley had a valley location too (ibid.).

But the comparison ends there. Unlike the Santa Clara Valley, Berlin in 1900 can hardly be said to have exerted a pull of amenity. It was, in the title of the most celebrated book about the capital of the Reich, 'the greatest rental barracks city of the world' (Hegemann 1930, title page): one of the most densely built and overcrowded cities to be found anywhere, with horrifying problems of poor housing and poor public health (Hall 1984). Whatever the other forces that drew high-technology industry to Berlin, its physical qualities were not among them.

Nothing is clearer, therefore, than that the Berlin electrical industry constituted a single, rapidly growing agglomerative complex. In this, a number of key factors played a continuing rôle. For sections of the industry dependent on raw materials (cables, transformers, heavy machinery) Berlin was not an optimal location – though no other inland site was. But much of the industry was not materials oriented (Czada 1969). Labour supplies, particularly of the skilled variety, were crucial, and Berlin had a unique advantage here in its supply of traditional handwork and artisan skills (Baar 1966). It also had plentiful supplies of cheap female labour for the more routine assembly activities (ibid.). Above all, the major market was here in the form of the public contractors: the railways, the post, the military. In 1925, it was estimated that one quarter of the production of the industry was sold directly in Berlin itself (ibid.).

The critical importance of this factor is seen throughout the industry's history. Siemens and Halske located their first and second workshops within minutes of the offices of the Post Office and the War Ministry, where they could obtain close contacts with officials (Czada 1969). Later, as competing firms entered (first in telegraphs, then in telephones) they too had to be close (ibid.). One, Mix & Genest, was founded in 1879 by a salesman, Wilhelm Mix, and a railway engineer, Werner Genest. It expanded rapidly to employ some 2250 workers by 1904. As it did so, it made a number of moves – but always within the Kreuzberg area, within a few minutes of the General Post Office. In 1905 the firm consolidated all its operations in Schöneberg, the next area outwards from the centre: a classic case of space-seeking (Anon. 1954).

All these firms, then, faced a conflict in their locational strategies: on the one hand they wanted close day-to-day access with the Reich ministries and agencies; on the other, they soon ran out of space and were forced further out. Abundant evidence shows how close and how complex were the ties between business and government. Siemens in 1879 had a serious lack of work in its railway signalling business: the railways, uncertain about the prospect of nationalization, would not place orders (von Winterfeld 1913). At the same time they fought hard to keep their monopoly in the supply of telephone equipment, which the Post Office was determined to break in order to lower prices (Peschke 1981). Later the General Staff encouraged the development of the Deutsche Betriebs-Gesellschaft für drahtlose Telegraphie (Telefunken) and were willing to provide subsidies because of their fear of an American monopoly (ibid.). Yet, having done this, they ordered equipment from Siemens's main competitor, Lorenz; for security reasons they wished to keep production in a few hands, but not to face a monopoly (ibid.). Finally, concerned with large state orders for equipment, the Reich bureaucrats had no coherent strategy for building up consumer-goods production for a mass household market; here Germany was already losing to the USA before World War I (ibid.). Thus their impact was sometimes positive, sometimes negative (ibid.).

Finally, the industry was locked into Berlin because it grew by the classic process of spin-off from parent firms, the same process that has been noted for Silicon Valley in the 1950s (Rogers & Larsen 1984). In the 25 years from 1903 to 1928, there were 12 recorded spin-offs from Telefunken; all except one established themselves in Berlin (Table 8.4) (Czada 1969). Undoubtedly, this was due in part to the continuing need for contacts with the Post Office; sometimes these were cemented through formal ties, as when the Lorenz-Werke, which formally broke from Telefunken, appointed as board members the State Secretary and Radio Commissioner and the President of the State Telegraph Office (ibid.). But there were also agglomeration economies in the supply of specialized equipment,

Table 8.4 Spinoffs from Telefunken, 1903–28.

Name	Period with Telefunken	Location of own firm
Hans Boas	with AEG before foundation	Berlin
Adolf Koepsel	*c.*1903	Berlin
Georg Seibt	1903	Berlin
Karl Kunsch	1903–06	Berlin
Erich F. Huth	1905	Berlin
David Loewe	1904–5	Berlin
Sigmund Loewe	1905–15	Berlin
Richard Heilbrun	1906	Nowawes near Berlin
Gustav Tropitz	1907–23	Cologne
Salomon Subkis	1910–14	Berlin
Walter Ritscher	1911–21	Berlin
Eugen Reiss	1911–18	Berlin
Joseph Massolle	1914–16	Berlin

Source: Czada (1969), based on *25 Jahre Telefunken: Festschrift der Telefunken-Gesellschaft 1903–1928*, 295–320. Berlin 1928.

and the knowledge of technical progress generally (ibid.).

This last was indeed a significant factor for the industry. Werner Siemens was in no doubt that organized scientific research was the basis of competitive power; he said in 1883:

> The industry of a country will never attain a leading international position, nor keep it, if it does not at the same time keep at the peak of scientific advance. (quoted in von Weiher & Goetzeler 1981, 43)

Werner Siemens had taken a leading rôle, in association with the Chief Postmaster, Stephan, in setting up the Deutsche Elektrotechnische Verein (German Electrical Union), a society concerned mainly with scientific matters, in 1879; he then campaigned for the establishment of a high-level national institute for fundamental research, which saw light as the Physikalisch-Technischen Reichsanstalt. His son Wilhelm campaigned with equal energy for a national scientific society akin to Britain's Royal Society, which resulted in the formation of the Kaiser-Wilhelm-Gesellschaft (Peschke 1981). Despite this, the rôle of research at Siemens remained ambiguous. Though separate research laboratories were set up for each of the firm's main activities, works managers were often hostile to them. Only in 1914, after the appointment of a new chief of research, did the company start work on a central research laboratory; only in 1919 was it completed (Siemens 1957b). But by the 1930s the Central Laboratory for Communications Engineering rivalled the facilities of the American Bell Laboratories (ibid.).

In this way, a firm like Siemens could internalize its externalities – but only up to a point. Even it depended on a climate of scientific advance in the world immediately outside. For smaller firms these contacts were even more crucial. Many industrial laboratories were established close to the Technische Hochschule in Charlottenburg (Treue 1976). Leading technical personnel at Siemens all came from an academic background, and there were frequent interchanges with the Hochscule (Peschke 1981). The general director of Mix & Genest moved to a new Chair of Communications Technology (Fernmeldetechnik) at the Hochschule in 1910. Though the Technische Hochschule in Berlin had been by no means the first to be founded in German-speaking Europe – the earliest, between 1806 and 1825, had been Prague, Vienna and Karlsruhe (Manegold 1970, Treue 1970) – its origins went back to the Berlin Technical Institute of 1821, founded by Peter Beuth, the Head of the Department of Trade and Industry in the Prussian Ministry of Finance, who had pioneered the cause of technical education in the service of industry (Henderson 1958); and inevitably, given the patronage of the Prussian and then the German state, it had become perhaps the most prestigious of all German technical institutions.

Scientific research, then, was a crucial factor in the clustering of high-technology industry in Berlin. But another critical element was the availability of venture capital. The growing industry needed prodigious amounts of capital; in one year, 1899, it was estimated that 212 million RM were invested in it, more than in any other industry – and this in an area that was still relatively experimental (Schulz-Hanssen 1970). Larger companies such as Siemens could finance most of their growth from profits (Baar 1966), but for other companies this was not the case. The available capital was extraordinarily concentrated in Berlin, which as early as the 1850s and 1860s, despite conservative Prussian bureaucratic controls on bank expansion and joint stock companies, had become the chief centre of the capital and gold markets (ibid.). By 1912 the big six Berlin banks had representation on the boards of some 751 companies (Zimm 1959). In this way they could keep control over their investments in what were often risky high-technology businesses.

The fall of Elektropolis

Berlin's electrical complex withstood the shocks of post-1919 reparations, the great inflation of 1923–4 and the Great Depression of 1929–32, though the last reduced output and employment by as much as 40% (Ch. 5). It received a boost from Nazi investment policies after 1933 and from rearmament after 1936 (Siemens 1957b). Indeed, in absolute terms – though it had long since relinquished first place to the Americans – it reached peak levels of both output and employment during World War II. Then, disaster struck.

The Berlin factories were repeatedly hit in air attacks during 1943–5, but generally managed to maintain production; thus the Mix & Genest plant was virtually destroyed in 1943 but was promptly rebuilt (Anon. 1954). Immediately at war's end the city was occupied by the Russians, who pursued an immediate policy of dismantling and shipping the plant. At Siemensstadt, of some 24 000 machines only 1300 were left, and only 400 were usable (von Weiher & Goetzeler 1981). Then, the firm lost its patent and market rights without compensation, and, for a time, was even forbidden to engage in research (ibid.). The firm found it easier, when the opportunity at last arose, to rebuild in West Germany. Already, impelled by strategic considerations towards the end of the war, it had created divisions there. It based its new structure on these, with bases in München, Nürnberg-Erlangen and Mühlheim-Düsseldorf. So, though the firm retained a presence at Siemensstadt, its base inevitably moved out of the city. Mix & Genest opened in an existing factory in Stuttgart in 1946, and moved its headquarters there in 1948; a critical factor was the fact that the Bundespost and Bundesbahn had moved their headquarters to Darmstadt and Frankfurt-Offenbach (Anon. 1954). AEG was in an even more unfortunate position, since several of its plants were in East Berlin and were nationalized.

AEG's problems illustrate, albeit in extreme form, the difficulties of continuing to operate in the old capital city. Firstly, the war destroyed an estimated 50% of the city's industrial plant. At the end of the war two-thirds of the remaining plant in West Berlin, one-third of that in East Berlin, was removed by the Russians; by 1947 West Berlin, with nearly two-thirds of prewar industrial capacity, accounted for less than one half the city's output. Then, reinvestment was inhibited by uncertainty about the political future, especially after the blockade and airlift of 1948. In Berlin as a whole, employment declined from nearly 790 000 in 1939 to

456 000 in 1959, or no less than 42%; in West Berlin the loss was 43%, in East Berlin 41%; and the loss in West Berlin was particularly marked in capital goods including electrical products (Storbeck 1964).

This, coming precisely at the starting point of the fourth Kondratieff upswing, proved lethal. The German electrical industry rebuilt itself, but, as we shall see in Chapter 10, it was never the same force again. And, as will be seen in Chapter 13, it rebuilt well away from Berlin. Effectively, and dramatically, this was the end of Elektropolis. The final symbolic act came only in 1984: AEG gave up production at its historic Ackerstrasse and Brunnenstrasse sites, and, by an appropriate irony, they were taken over by the Technische Universität, the old Hochschule, as a technology transfer centre (Allesch 1984).

Boston to New York: the emerging electrical Megalopolis

During the mid-1870s Berlin was quite suddenly faced with the emergence of a rival electrical and information-technology metropolis. The New York region, and in particular Thomas Edison's Menlo Park laboratory, became a hive of innovation, producing in quick succession the automatic multiplex telegraph, the gramophone, and the incandescent electric lamp. American electrical and information technology was thus uniquely associated with the innovatory genius of one man. The question is why he chose the New York region as a place to innovate. Edison, who in his lifetime filed no less than 1093 patents, was born in the small midwestern town of Milan, Ohio. He migrated to Boston at the age of 22, to work for Western Union, because it was then the unique centre of scientific and electrical research in the USA (Josephson 1961, Wachhorst 1981). A biographer writes:

> Yet while Edison had been merely an oddball in the offices of the Midwest, in Boston he was one of numerous young men bending their imaginations to invention. New England was the cradle of the American Industrial Revolution, and the city was the crucible of scientific and especially electrical research. (Conot 1979, 28).

Charles Williams's electrical equipment shop, where many young men rented small rooms as laboratories, was the fulcrum of electrical experiment in the USA; Edison went there to buy equipment and exchange ideas, drawing on the ideas of others. Six years later, Alexander Graham Bell would rent a room in the same neighbourhood (ibid.). The Massachusetts Institute of Technology had admitted its first students in 1865 and had developed a physics laboratory – out of which electrical engineering was later to grow – the year after (Loria 1984).

But, resigning to become a freelance inventor, Edison soon moved to New York. The reason was that, sensitive to the demands of the market even then, he appreciated the acute needs of Wall Street's financial community for better information technology. Introduced to Frank L. Pope, one of the top experts in telegraphy and supervisor of Laws Reporting Telegraph in New York, he was in turn introduced to its proprietor Samuel S. Laws, an early indicator of gold prices. Edison, realizing that the system required a printer, began to work on this, then on stock tickers for Western Union (Josephson 1961). He set up partnership with Pope, who 'had entree to most of the business world concentrated on Wall Street and Lower Broadway' (Conot 1979, 34) in 1869. Since he was lodging with Pope's parents in Elizabeth, New Jersey, across the Hudson from Lower Manhattan, he found a place in an old electrical factory in Jersey City, then in neighbouring Newark (Conot 1979, Josephson 1961); cheap premises may also have been an attraction. He obtained venture capital from Philadelphia financiers through his New York contacts (Conot 1979). Edison's Universal Stock Printer, developed for (and funded by) Western Union, was soon adopted by exchanges throughout the USA and Europe (Josephson 1961); constantly improved, it was virtually the prototype of all new information technology. Then, through the mid-1870s, working for both Western Union and its chief rivals the Automatic Telegraph Company, Edison worked to develop an improved automatic telegraph machine, finally dispensing with the need for morse code on the part of the operator and providing automatic receipt and recording and a multiplex telegraph to increase the capacity of existing connections (ibid.).

In 1876 he made a dramatic move (Fig. 8.2): abandoning his production activities, he moved to a new research laboratory at Menlo Park, New Jersey, an isolated hamlet 12 miles south of Newark and 25 miles south-west of New York City: 'the first industrial research laboratory in America, or in the world' (ibid., 133). Here land was cheap because of a depression in agriculture; despite its rural location, it was easily accessible to New York via the Pennsylvania Railroad (Conot 1979). Deliberately led by the market, Menlo Park aimed at 'a minor invention every ten days and a big thing every six months or so' (Josephson 1961, 133–4). Here, the phonograph and the incandescent light were developed. New

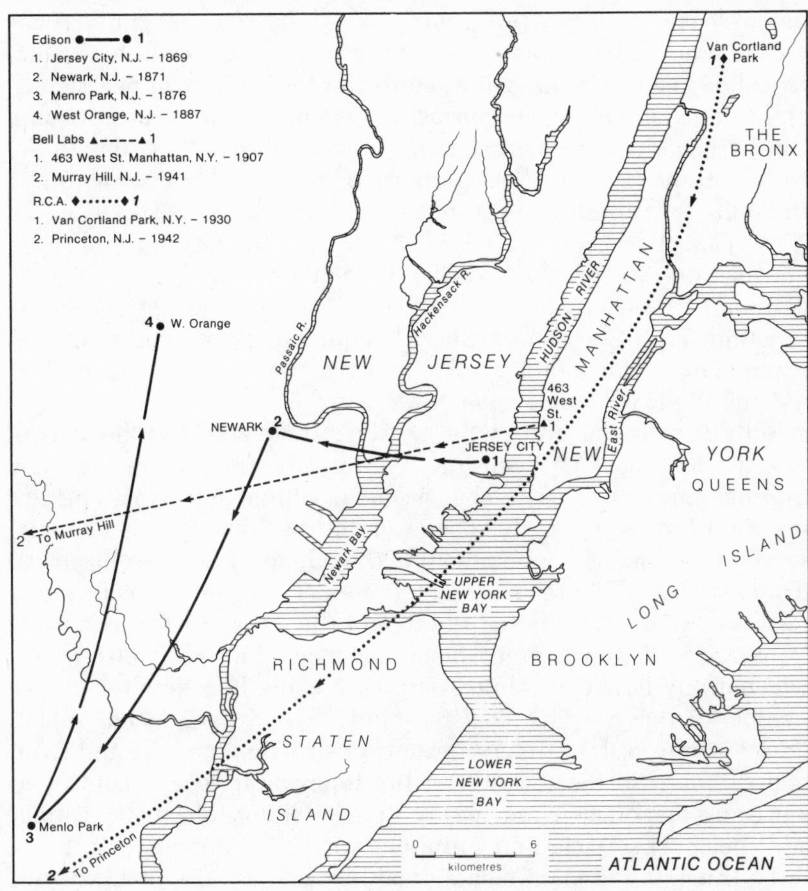

Fig. 8.2 New York City region: electrical industry, key locations, 1869–1942

York City capital, with the leading investment house of Drexel, Morgan & Company in a central rôle, financed Edison's expanding activities at this time (Passer 1953, Chandler 1975). A new laboratory at West Orange, some 10 miles away, was opened in 1887 and was the largest, best equipped in the world (Josephson 1961, Conot 1979). Meanwhile, Edison had gone back into production; in 1886–7, following a labour dispute at the unionized New York factory, he moved the facility to Schenectady 200 miles from New York (Josephson 1961).

Edison's location decisions were clearly rational. It made sense, in an innovatory industry then experiencing a quite extraordinarily intense burst of innovation, to concentrate on R&D. Equally, it made sense to locate this function in what one biographer has called 'a monastery where he could devote a cloistered life to what he called "communion" with nature and the probing of her secrets' and to escape the 'vulgar' curiosity of visitors to the Newark shop. But he nevertheless took pains to publicize his innovations to the New York press – impelled, perhaps, by his lawyer who saw this as the route to Wall Street venture capital (Wachhorst 1981, 38–9, 41). The Menlo Park site – just far enough from the city for seclusion, yet on the main New York–Philadelphia rail line and thus accessible to the city for publicity and finance – was optimal for its function. To it Edison attracted a unique bunch of highly talented technicians, who would probably have been impossible to find outside the New York region.

Both directly and indirectly, Edison's innovations at the start of the third Kondratieff can be seen as the foundation of new information technology. The ticker machine and stock printer responded directly to the needs of the fast-developing financial services sector. The phonograph, though for long non-electrical (Ch. 7), was 'the precursor of all modern forms of mechanical memory' (ibid., 22); the light bulb and the 'Edison effect' were the forerunners of the vacuum tube or valve, on which early·electronics was entirely based (Wachhorst 1981). For the first time, machines (which in the previous, palaeotechnic, era had replaced human muscle) now began to serve extensions of the eye, ear and even brain (ibid.). So the location of the innovation centre, controlled partly by market demand, partly by availability of labour, capital and specialist services and suppliers, was of the utmost importance – the fringe of the giant commercial metropolis. After merger with Thomson-Houston to form General Electric in 1891, the headquarters remained in Manhattan. But long before this R&D had been hived off to the rural fringe; and, as soon as long production runs developed for a secure and expanding market, these were decentralized also (Hund 1959). This model, of course, was to recur subsequently in the development of Highway 128 and Silicon Valley.

Another key figure, Alexander Graham Bell, had also started his inventive career in Boston, coming there in 1871 on what was at first a short contract to teach at the Boston School for Deaf Mutes. Almost immediately, he was tapping the city's vast stock of scientific and technical knowledge, including that of the Institute of Technology (Bruce 1973). Charles Williams's workshop, also a magnet for Edison, made the early instrumentation for his assistant Thomas Watson, and was later to become a nucleus of Western Electric (Fagen 1975). In Boston, too. Bell found local venture capital in the form of his future father-in-law, Gardiner Greene Hubbard, and Thomas Sanders (Bruce 1973). After the technical breakthrough and the start of commercial business, it was a group of Boston businessmen who provided the capital and who remained in effective charge until 1907 (Fagen 1975). Nevertheless, AT&T and Western Electric established their headquarters at 463 West Street in New York City in 1897; and it became the parent of the Bell Corporation in 1899, because New York's tax laws were more favourable than those of Massachusetts (ibid.).

For a time the research departments of AT&T and Western Electric remained separate, the first in Boston, the second in Chicago. But in 1907, when for economy reasons they were reorganized under one roof and one director, they were brought together in the New York headquarters. Here the operation grew from a few people to a total of 3600 by 1925. In that year it became Bell Laboratories, a separate organization responsible to both AT&T and Western Electric; the New York headquarters was extended. Down to World War II, it handled a bigger annual research budget than any American university (Fagen 1975, Brooks 1976, Noble 1977, Reich 1985). Finally, in 1941, Bell Laboratories beat the path that Edison had followed three-quarters of a century earlier, moving to Murray Hill, New Jersey (Brooks 1975), only a few miles from Edison's old West Orange laboratory in New Jersey.

Long before that, in the last decade of the 19th century, Edison's example had attracted innovative competitors like Nikola Tesla, A.E. Kennelly and Reginald Fessenden. Thus a regional agglomerative effect began to occur, similar to that later seen in Highway 128 or Silicon Valley. Ironically, at this point Edison's research tradition deserted the New York region. General Electric, the company formed out of the Edison and Thomson-Houston empires in 1892, at first neglected research. When the company reversed the policy in 1900, it seemed logical to put the new laboratories at the manufacturing headquarters in Schenectady, the location Edison had chosen in 1886 to avoid the New York City unions and to obtain more space. Willis Whitney, the MIT professor who left academia to direct the new operation, built up an outstanding operation in both pure and applied research, with over 300 employees, by the mid-1920s (Noble 1977, Reich 1985, Wise 1985).

The next influence, somewhat unusually, came from abroad. Gugleilmo Marconi's success in transmitting Morse code by wireless telegraphy across the Atlantic, in 1901, logically soon led to the establishment of a Marconi subsidiary in New York, since here was the obvious place for transatlantic traffic, and the largest mass of population and business to provide markets. This company already hired Columbia University's electrical engineering professor Michael Pupin as consultant – an early example of a university–industry link.

But, by 1910, American Marconi was experiencing strong competition from General Electric, operating both within and outside the New York region, and AT&T. The latter, increasingly interested in the electronic valve, bought the many patents of the American Lee De Forest, inventor of the triode valve (Hund 1959). But GE's brilliant Irving Langmuir in effect produced a new version of the tube, leading to a patent battle that lasted until 1925 (Wise 1985, Reich 1985). World War I provided a tremendous boost to the development of the industry, especially through demand from the navy; and, despite advances from the laboratories of the American firms, the largest source of supply remained the Marconi factories in Aldene and Roselle Park, New Jersey. This spurred Franklin Delano Roosevelt, then Assistant Secretary to the Navy, to urge American industrialists to co-operate in creating an indigenous competitor.

The birth of the Radio Corporation of America (RCA), in New York City in 1919, has been described as 'probably the single most important factor in shaping the structure of the electronics industry' (Hund 1959, 253–4). It absorbed the assets of American Marconi and acted as a patent pool for General Electric, AT&T, and Westinghouse (Archer 1938, Maclaurin 1949, Noble 1977, Reich 1985, Wise 1985); it established R&D facilities (mainly the latter) at van Courtlandt Park in New York City, opening a separate research establishment at Princeton in 1942; it acquired radio stations, and acted as a sales organization for the products of its member firms, only starting to manufacture itself much later (Maclaurin 1949). From 1920 the radio industry rapidly grew – not only in New York, the leading centre, but also Chicago, Boston and Philadelphia. By the start of World War II it was producing 13 million sets a year (Hund 1959). Meanwhile scientists were working on the development of television, especially at Westinghouse before 1930 and at RCA afterwards. By 1931 a trade paper was announcing that television 'may be the looked-for new industry to bring the army of the unemployed out of the trenches by another Christmas' (ibid., 255). It proved prophetic, though the timing was wrong: the fourth Kondratieff was not to take off for another two decades.

During this period, the New York region lost some of its earlier dominance. In consumer electronics it took an early lead in radio, but was overhauled by Chicago as early as 1929; the same story happened in the late 1940s with television. So by 1954 the Midwest had become the centre of the radio and television industry, with nearly 35% of production against New York's 15% (Hund 1959). New York's early gain reflected the early concentration of both R&D and market demand there, with many infant firms; but, as the size of the market grew and the technology stabilized in the late 1920s, there was a strong incentive to concentration. In 1932, as the result of a consent decree in an antitrust settlement, RCA was to become a direct manufacturer and General Electric and Westinghouse were to refrain from production for two and a half years; this fostered the growth of some other survivor firms – Philco, Emerson, Zenith, Galvin (Motorola) – which became household names. Philco, as its name suggests, was Philadelphia based; the others mainly developed in Chicago (Maclaurin 1949).

By the 1950s, then, about ten firms produced at least 80% of output. Only one of them, Emerson Radio and Phonograph, was consistently New York based. Exactly the same process of consolidation occurred with television between 1947 and 1957, when 10 major producers emerged – two, Emerson and Westinghouse, in the New York region, four in the Chicago area (Hund 1959). Hund, studying the industry at the end of the 1950s, concluded that the crucial factor in Chicago's dominance was its central position for supplying the national market, a factor enhanced by the spread of truck transportation in the 1930s and 1940s (ibid.). Similarly, as electronics invaded the phonograph in the new hi-fi industry immediately after World War II, the industry at first concentrated in small firms in the New York area, then again dispersed to the consumer electronics strongholds of the Midwest (ibid.). And, in so far as consumer electronics remained in the New York region, it increasingly moved from small Manhattan workshops to New Jersey factories – a move propelled in large part by advantages in shipping cost (ibid.).

Also, in the supply of components such as valves New York had a commanding original lead, and for the same reasons – the importance of R&D, the small size of the firm, the importance of face-to- face contact. Valve manufacture, in particular, was concentrated in and around Newark in a number of RCA plants, located here because the skills were available in Edison's electric lamp factories (ibid.). But, though RCA continued to dominate valve production, a number of smaller competitors (Sylvania, Raytheon, National Union) won an increasing share during the 1930s. One, Sylvania, competed effectively by low-cost production in small towns in Massachusetts and Pennsylvania; Raytheon, another Massachusetts-based manufacturer with an origin in electrical map manufacture, made pirated tubes (Maclaurin 1949, Estall 1966).

During and after World War II, these firms grew mightily. By the

1950s New York firms accounted for only a minority of production (Hund 1959). Much of the production of capacitors and resistors, too, had moved outside the region, impelled by the search for cheaper labour as the market grew and the technology stabilized (ibid.). By the 1950s, the one branch of electronics where New York still had a commanding position was military and industrial electronics, where the prevailing market uncertainty, the unstandardized nature of the product, and the heavy reliance on engineering talent and skilled labour gave a premium to the very large metropolitan area (ibid.). But, even then, new areas of specialized electronics production, such as the San Francisco area, not yet christened Silicon Valley, were starting to appear. And this is a fourth Kondratieff story.

By the end of the third Kondratieff and the start of the fourth, then, New York's electronic dominance was already much weakened and was weakening further year by year, albeit disguised by rapid expansion of this group of industries. In 1947 the metropolitan area had 16% of national employment in radios and related products, 35% in electronic tubes (including valves) and 45% in scientific instruments; corresponding percentages for 1954 were down to 15, 16 and 39 (ibid.). A New York location, Hurd concluded, was a 'passing phase'; it would persist only for those branches in which uncertainty – about technology, about demand – dominated the scene (ibid.). Thus, without the experience of wartime destruction or defeat, New York like Berlin was steadily losing its status as Electropolis.

A strange combination of circumstances favoured its old rival, Boston. From at least the 1860s, when Edison and Bell migrated there, Greater Boston had remained indisputably the pre-eminent centre of scientific and technological research on the East Coast, and even – at least until the 1950s – in the entire USA. During the 1920s and 1930s MIT's Department of Electrical Engineering, founded as early as 1882, built an unchallenged position in electrical and electronics research. Here occurred Bush's pioneer work on the analogue computer, and Forrester's postwar construction of one of the first programmable digital computers; here also was the scene of wartime advances on radar (Noble 1977, Wildes & Lindgren 1985). This work, which developed links with Raytheon as a local supplier of equipment, was a critically important basis for the dramatic growth of the area's electronics industry in the 1950s, which we describe in Chapter 13.

This points to an important conclusion. At the start of the electrical age Boston was the centre of scientific invention, New York the centre of commercial innovation; Edison, in moving from one to the other, was deliberately following the pull of the market. Thence, with the institutionalization of research in Bell and GE laboratories after 1900 and improvements in communications and transport technologies, the specific locational advantages of the giant commercial metropolis became irrelevant: Bell and RCA research happened to remain there until 1941, GE located in Schenectady, and all three operated effectively. All these laboratories were doing fundamentally important work in the 1930s – Zworykin's research on television at RCA, Shockley's on the transistor at Bell, for instance. Yet many of the new and successful firms in the radio industry were imitative rather than innovative, and just after World War II Maclaurin could conclude that the industry lacked a scientific entrepreneur who could apply the latest scientific advances – a quality that would be vital for the future (Maclaurin 1949). In the event, that quality would emerge, not in radio but in the newly emerging information technologies of the 1950s. And it would manifest itself first in Massachusetts, then in California. This is a fourth Kondratieff story, to be told in Chapter 13.

The comparative case of London

In the third Kondratieff the UK steadily fell behind Germany and the USA in the electrical and electronics race; London was no Berlin, no New York. At the start of the 20th century, electrically speaking, 'London was a backward metropolis. Electric power was believed to be the hallmark of progress in enlightened cities like Chicago and Berlin, and London had notably little of it' (Hughes 1983, 227). One basic problem was the disordered, fragmented nature of London's electrical supply industries, with 65 different undertakings, 49 distinct supply systems, 32 different voltage levels for transmission and 24 for distribution, and 70 different systems for charging and pricing (Hughes 1983). In 1913 each Londoner consumed only two-thirds as much electricity as the average Berliner, only just over a third as much as the average Chicagoan (ibid.).

Yet the UK did remain the world's third largest producer of electrical equipment, and its industry was just as strongly concentrated in London as Germany's was in Berlin, or, originally, the USA's was in New York. Thirty-seven per cent of all electrical engineering workers were found there at the census of 1921, the first point at which we have reliable figures on employment by workplace; the proportion was 34% just after World War II, in 1951 (Hall 1962).

The earliest branch to develop, as in Germany and the USA, was telegraphy. The UK, as the world's leading maritime power, was in a unique position to exploit the booming market for international telegraph links that developed in the 1850s and 1860s. The Atlantic was spanned, in the British-built Great Eastern, in 1858. Logically, since the product was destined for the ocean bed, this branch located on navigable water where the cable could be taken straight from the holds of factories and coiled in the holds of ships (Smith 1933). The materials like the product were heavy, and were often imported by water too – copper, lead, iron, rubber, paper, cotton, oils, bitumen and other pitch substances (COIT 1928b). The plant itself was heavy and needed space.

Logically, on all these grounds the industry established itself on Thamesside sites at the then edge of London, where land was cheap. In the 1890s cable making was found in a few large factories in Woolwich on the south bank, and at Silvertown opposite (Duckworth 1895). It was still there 40 years later: in 1933, out of 19 members of the British Cable Makers' Association, 9 were in London and 7 of those were on the riverside downstream from Tower Bridge (Smith 1933). Siemens established its British cable manufacturing base at Charlton in 1863; frustrated in 1899 by building regulations in its attempt to add an electrical machinery works there, on the model of its giant Charlottenburg works, it established the base at Stafford in the Midlands (Martin 1966). The innovations of the third Kondratieff created other heavy branches of the industry (switchgear, electrical control apparatus, and the like), which sought similar locations. The British Electrical Transformer Company chose a site at Hayes, then deep in the open countryside west of London, in 1901 on the banks of the Grand Union Canal and next to the main line of the Great Western Railway (ibid.).

But the more characteristic branches of electrical engineering in London – and the more significant for the later development of the new information technologies – were of quite a different character. Just as in Berlin, they arose as small workshop plants in the heart of the inner city, making specialized high-value products. Certain areas, among which Clerkenwell was outstanding, played a crucial rôle. From the start of the 19th century Hatton Garden was a centre for jewellery and precious metals; thus the infant electrical and radio industries could readily buy precious metal contacts and contact by-metal there (ibid.). Here too was a unique concentration of invention and experiment, making this the centre for precision instruments of all kinds. Ferranti in 1883, aged only 19, began his business first in Hatton Garden, later in Charterhouse Square in the heart of Clerkenwell (ibid.). Similarly, scientific glassware proved an important basis for the start of the electronic industries; A.C. Cossor established his small workshop in Farringdon Road, Clerkenwell, in 1896, making a great variety of such glassware including early fluorescent tubes, the first British cathode ray tube (in 1902); by World War I it had turned to wireless transmitters and valves, staying with these products for the civilian market after 1918.

Some of these firms might remain in the inner city for a long time; others were forced out in search of space, as for instance Cossor who moved to Highbury in 1922 (ibid.) This was typical; firms moved to a wide crescent north of Clerkenwell, stretching from Camden Town to Dalston (ibid.) Siemens, too, took over a Dalston factory for the manufacture of tantalum lamps, a highly specialized and delicate task involving much individual skill and judgement (ibid.). Then, expansion took place even further afield. Edison Swan, establishing its carbon lamp factory at Ponders End before World War I, was a pioneer.

Yet another origin of these precision trades lay to the south and south-west of London's West End, in Pimlico and Lambeth. Both were on marshy sites close to the centre, offering cheap sites and water-borne raw materials. The south bank of the Thames, from Vauxhall Bridge to London Bridge, was in the early 19th century the greatest concentration of engineering shops, large and small; Pimlico, on the opposite bank, had the additional advantage of closeness to government, a feature, as already noted, that it shared with Berlin's Kreuzberg. Thus James Simpson began here in 1790, making pumps and fire engines for the government, and a century later the factory was still in the area; Siemens began a British operation here in 1858 (ibid.). Towards the end of the 19th century, firms began to move out from this area, too, in search of space; and the new electrical industry established itself close by, in places like Hammersmith, where it could draw on the concentration of engineering skills and on large supplies of available female labour. Additionally, in industries like electric lamp making, the early demand was very highly concentrated in London; and the products were fragile and bulky (ibid.).

Thus, by the beginning of the 20th century, the London electrical industry, like its Berlin counterpart, was establishing itself on peripheral sites. As well as the traditional centre for cable making in Thamesside, East London, two major peripheral locations were evident: one, an outgrowth of Clerkenwell, in the marshy Lea Valley of north-east London; the other, spreading out from Pimlico and Lambeth, in the Hammersmith and Acton areas of west

London. Hammersmith in particular included two electric lamp factories, and during World War I these naturally diversified into radio-valve manufacture, and new firms, such as the then small-scale business of Captain S.R. Mullard, working for Admiralty contracts, joined them in the same area (ibid.).

Transport links may well have played a rôle here: Hammersmith was on the District Line of the Underground, which ran out from Pimlico; the Lea Valley was on a direct line of rail from Liverpool Street station, with cheap workmen's trains. But they proved even more significant between the two world wars, when the growth of both capital and consumer electrical goods caused huge expansion of these industries. Between 1921 and 1951 employment in the electrical industries in London grew from 51 000 to 183 000, or three and a half times. Firms moved out, and new firms joined them, mainly along radial lines from the existing concentrations. Thus the pioneer firms in the Lea Valley were joined by many others, and the west London concentration spread out from Hammersmith and Acton into a wide sector embracing Park Royal, Hayes and Southall, the Great West Road and Slough. Particularly notable in this was the rôle of the new arterial by-pass roads such as the Great Cambridge Road in the Lea Valley, or the Great West Road and Western Avenue in West London. They carried both the materials and components and the finished product; often, too, they served as routes for the buses which brought the workers. D.H. Smith's pioneer study of the industries of Greater London, published in 1933, showed a total of 140 000 workers in north and north-west London, of whom no less than 75 000 were in the western sector (Smith 1933). Not all of these were in the high-technology industries of their day, let alone information technology. But 8000 were in electrical equipment and another 6000 of them in gramophone manufacture, which during and after World War II was to prove a particularly potent source of information technology spin-off (ibid.).

London's complex of electrical industries was not destroyed by World War II, as was Berlin's. But the war did begin a dispersal, for strategic reasons, which continued with accelerated force afterwards, as the infant electronic industries, growing directly out of the radio, gramophone and lighting firms of the 1930s, moved out of London in search of space. By then the Green Belt was placing real limits on their locational choice. Less dramatically than in the case of Berlin – for they were constrained not by Russian occupation and the threat of nationalization, but by the edicts of the town and country planners – they began to migrate from the metropolis to sites 30, 40, even 60 miles distant. But this is a story that will be taken up where it belongs, in Chapter 13.

PART FOUR

The Electronic age

New IT in the fourth Kondratieff, 1948–2003

9
The electronic revolution

Electronics represented a product of the third Kondratieff, but (as seen in Ch. 6) a late one: though the basic innovations came in the first decade of that long wave, their mass industrial applications – in radio in the 1920s and 1930s and in radar during the 1940s – coincided with its downswing. It was thus only at the onset of the fourth Kondratieff, during the early 1950s, that the pace of electronic innovation dramatically quickened, starting to create a stream of microelectronic components and systems that became the core of the modern New IT. And this in turn became a key technology of the fourth Kondratieff, as distinctive in its turn as cars and electrical products and pharmaceuticals in the third, Bessemer steel and machine tools in the second.

At last New IT was free of the constraints of crude mechanical systems, which had aborted the birth of Babbage's computer and hindered the early development of many office technologies. Information, stored in huge amounts and processed by increasingly miniaturized systems, progressively became the main agent of economic growth; during the fourth long wave, the main emphasis in the world's advanced industrial economies was moving decisively from the production of goods to the manipulation of information: and there occurred a shift towards the 'convergence of all modes of communication and information technology into one common digital art of transmission, processing, storage, and retrieval, whatever the content, whether it be sound, sight, figures or symbols' (Inose & Pierce 1984, 1).

The transistor: core technology

The invention of the transistor (at Bell Laboratories, Murray Hill, New Jersey on 23 December, 1947) was without doubt one of the technological keys that opened the fourth Kondratieff. The transistor is a device which uses semiconductor materials to perform functions such as amplifying or switching an electrical current. It has many advantages over the thermionic valve on which the electronics industries were based during their first half-century: significant reductions in size and energy consumption, increased reliability and range of applications. Its invention was the work of three physicists at Bell, John Bardeen, Walter Brattain and William Shockley, but it stemmed also from the endeavours of many hundreds of researchers working in many different places on related electronic technologies during and before World War II. Down to 1941, indeed, the UK and perhaps Germany were ahead in the field, but then – with a massive wartime effort in radar detection at MIT and other places – the USA took a clear lead. Yet its discovery in the Bell Laboratories was far from accidental; it was a direct result of the huge state-supported scientific and technical resources dedicated to the work, in what was already the world's largest industrial research organization. But fundamental research in the UK may have lagged behind Bell by only a matter of weeks (Braun & Macdonald 1982).

The first Bell transistor, a germanium point contact transistor, was soon adopted for commercial production by Western Electric, the manufacturing arm of AT&T; but it proved very difficult to manufacture in bulk and suffered from many problems of reliability. Some of these difficulties were overcome in 1951 by Shockley's invention of the junction transistor. But commercial applications and interest in the transistor were rather slow in the early years of the 1950s, in part because of continuing difficulties in standardized mass production techniques. Indeed, process innovations proved almost as important as product innovations in the subsequent history of electronics components technologies (OECD 1968, Braun & Macdonald 1982).

One of the most crucial came in May 1954 when Texas Instruments succeeded in making the silicon transistor; it had the capacity to work at very high temperatures and proved of great interest to the military, whose support provided a major impetus to the development of transistor production in the USA over the following decade (OECD 1968, Braun & Macdonald 1978). A second was the planar process of 1959, which offered many advantages over previous methods for the standardized production of transistors and later more integrated and compact components. It came from Fairchild Semiconductor, a relatively new company started by eight scientists who had previously worked at Shockley's Palo Alto laboratories before starting up with support from Fairchild Camera and Instrument Corporation. By making semiconductors relatively robust and reliable, the planar process did much to set the industry on the growth path it followed for the next quarter-century; from 1959 to 1962 prices fell by 80–90%; between 1957 and 1965 production increased twentyfold by volume (Braun & Macdonald 1982, Dummer 1983).

The second key product innovation was the integrated circuit, which perhaps marks the real step towards the establishment of microelectronics as a major new technology system (Dosi 1984a, 1984b, Perez 1985). It allowed the manufacture of many individual electronic components and functions on a single microchip. It was conceived theoretically as early as 1952 by G.W. Dummer of Britain's Royal Radar Establishment (the RRE) (Freeman *et al.* 1982, Dummer 1983). The first working model was developed under contract to the RRE by the British firm of Plessey in 1957 (OECD 1968); the first patent, a rather crude device, was filed in 1959 by Texas Instruments. It was finally Robert Noyce at Fairchild who demonstrated how these devices could be easily and reliably manufactured using the planar process. The first commercial application of integrated circuits was in hearing aids in 1963 but the military provided the most important early market; by 1970, however, with early production problems resolved and prices lowered, their share was down to one-third (OECD 1968, Braun & Macdonald 1982). As with the transistor, the 1960s saw progressive process innovation, producing increasing numbers of electronic functions on a single microchip, with rapid falls in price per function plus increased reliability and range. The discrete components incorporated in each integrated circuit increased from less than 10 in 1960 to 100 000 in 1980, and the average price of the integrated circuit fell from $50 in 1962 to about $1 in 1971 (Braun & Macdonald 1982).

The third major product innovation was the 'microprogrammable computer on a chip', or microprocessor, developed by an Intel team headed by Ted Hoff in 1971. By the mid-1970s the industry had completely changed its emphasis from the manufacture of discrete integrated circuits to the manufacture of what were in effect small computers; about 40 different microprocessors were being produced in the USA by almost all the major semiconductor manufacturers, with full support systems available. The early 4-bit and 8-bit microprocessors were followed by 16-bit and 32-bit devices in the early 1980s, and the size of the standard dynamic random access memory chips has increased from 1k in the early 1970s to 64k and 256k by the mid-1980s (Braun & Macdonald 1982, OECD 1985).

It was the microprocessor with its capacity to perform 'the full range of logic functions on every kind of information presentable in digital form which is responsible for much of the rapid diffusion of modern electronics into new applications in new areas' (Braun & Macdonald 1982, 112). Aided by the new MOS (metal oxide semiconductor) technology, which became commercially viable only in the early 1970s, an explosion in capacity took place both in microprocessors and in memory, from 5 transistors per chip in 1962 to some 150 000 in 1982; by the 1980s one chip would carry as much information as a middle-range machine of a decade before, and at negligible cost. The main consequence was a further dramatic fall in the time and cost of processing information; the price per bit of dynamic random access memory fell at an average rate of about 35% per year from 1970 and is expected to continue at this rate into the future (Braun & Macdonald 1982, OECD 1985). The direct consequence, as will shortly be described, was that the computer became a cheap mass product.

These advances came almost exclusively from industry, albeit with major support from government and especially from the military, which both paid for R&D and provided much of the early market, above all in the USA; in the mid-1950s defence contracts accounted for one-third of all semiconductor sales, by the late 1950s one-half (OECD 1968, Braun & Macdonald 1982). At first the established valve firms were the main beneficiaries of military R&D contracts, winning 78% in 1959, but even then new firms had 69% of procurement contracts, allowing them to develop new products and thus pass rapidly down the 'learning curve' to high volumes and low costs (Braun & Macdonald 1982).

The result was the emergence and growth of many new firms, conforming to the Schumpeterian model of innovation; although many large established firms with high levels of R&D were responsible for about half the major innovations between 1950 and 1980, the large firms' share of the market showed little change, but the top positions in the industry showed kaleidoscopic changes as one technology succeeded another (Table 9.1). In the European semiconductor industry, in contrast, very few new small firms or 'spin-off' firms entered; the new technology was introduced almost exclusively by subsidiaries of American firms or by large indigenous firms – Philips, Siemens, GEC, AEI (OECD 1968, Tilton 1971, Sciberras 1977, Braun & Macdonald 1982, Freeman *et al.* 1982).

The birth of the computer

The electronic computer, although closely related and partly dependent upon advances in electronic component technologies, in

Table 9.1 Leading US semiconductor firms, 1950–79.

	Merchant manufacturers, by share of world market		
Valves *c*.1950	Transistors 1955	Semiconductors 1965	Integrated circuits 1979
1 RCA	1 Hughes	1 Texas Instruments	1 Texas Instruments
2 Sylvania	2 Transistron	2 Motorola	2 National Semiconductors
3 GE	3 Philco	3 Fairchild	3 Motorola
4 Raytheon	4 Sylvania	4 General Instruments	4 Intel
5 Westinghouse	5 Texas Instruments	5 GE	5 Fairchild

Source: Braun & Macdonald (1982).

many respects may be regarded as having 'the most universal and far-reaching consequences' of all 20th century technological developments (Margerison 1978).

As with many key inventions, there is some doubt and dispute among historians of technology about the origin of the computer. It appears that the first practical computer, the Z3, was produced by Zuse in Germany in 1941, and that by 1942 his Z4 model was being used for aircraft design calculations. But this was an electromechanical device, as was the Harvard Mark 1 (also known as the Automatic Sequence Controlled Calculator – ASCC), produced by a team led by Aiken between 1937 and 1943; it became fully operational in May 1944. In 1943 and 1944 Bell Laboratories in the USA also produced two important devices, and in 1944 a British team at Bletchley Park produced a number of computers which appear to have involved important technical advances (Freeman *et al.* 1965, Margerison 1978, Randell 1980). All these, however, were electromechanical, and by the late 1930s it was already clear that electronic computers would be much faster and potentially more flexible. In Germany Zuse and Schreyer developed a crude model in 1942, but soon afterwards government support was withdrawn (Freeman *et al.* 1965, Zuse 1980). In 1944 a British team working in secret produced the Colossus, one of the first externally programmed electronic digital computers (Randell 1980).

It seems generally agreed that the first general purpose electronic computer was the ENIAC, developed for the American army to calculate trajectories of shells and bombs, and completed in 1946 by a team led by Mauchly and Eckert at the University of Pennsylvania. Shortly after that, the critical step towards the modern computer was made with the design of a practical stored-program computer, the EDVAC (Randell 1980). Even these early electronic machines performed arithmetic functions approximately a thousand times faster than electromechanical ones.

From 1945 to 1955 great progress was made in solving some of the problems of logic design, programming techniques, storage systems and appropriate peripheral devices (Freeman *et al.* 1965, OECD 1969). By 1951, the UNIVAC 1, the first commercial electronic computer, had arrived in the USA; developed by Eckert and Mauchly, who had left the University of Pennsylvania to set up their own company (acquired in 1950 by Remington Rand), it could handle both numerical and alphabetical data. In the UK the most important work in these years took place in London, Manchester, and Cambridge universities and the National Physical Laboratory. In 1951 a British computer, based partly on Cambridge University technology and known as the LEO (Lyons Electronic Office), was the first to be used purely for commercial data handling (Margerison 1978).

Most of this work was heavily supported by government, especially for military purposes. Eckert and Mauchly's ENIAC was financed by the army; Forester's Whirlwind project at MIT, immediately after the war, by the navy (Brock 1975, Wildes & Lindgren 1985). Despite the advent of UNIVAC and LEO, by the early 1950s there was little commercial recognition of the potential of computers (Freeman *et al.* 1982). IBM's President Thomas Watson was still agnostic as to commercial prospects, and his own marketing department agreed; the success of UNIVAC, used to process the 1950 United States Census, took them by surprise, posing what seemed to be a major challenge by the rival Remington Rand Corporation (Sobel 1983). At this point IBM began in earnest to develop a computer to compete with the UNIVAC, releasing the 701 vacuum-tube machine in 1953. Already by 1955, IBM had a 56% market share against Remington Rand's 38%; by 1957 the shares were 78 and 16% (Brock 1975).

In 1958 the advent of 'second-generation' electronic computers utilizing transistors – pioneered by Sperry Rand, formed in 1955 by the merger of Remington Rand and Sperry, but emulated by IBM with their 7090 a year later – made clear the commercial potential of computers, and the industry began on a rapid growth path. This was marked in particular by the arrival in the mid-1960s of the minicomputer, with its combination of computing power and relatively low cost. New companies – Control Data, a breakaway from Rand, and Digital – began to emerge. Together with precomputer business machine firms, such as Sperry, Honeywell, Burroughs and NCR, that moved into the field by the early 1960s they formed the 'Seven Dwarfs' that competed with IBM (Brock 1975, Fishman 1981).

A similar technological step function took place in the mid-1970s with the introduction of the microprocessor; the large-scale integration of the components needed for the central processing unit of a computer onto a single microchip meant that the costs of computer hardware fell dramatically. These developments led to the emergence of an important new branch of the computer industry from the late 1970s, accompanied by its own process of swarming: the microcomputer (personal and home computer) industry. The beginnings came as early as 1975, when MITS, a small firm based in Albuquerque New Mexico, announced the Altair, a primitive machine, 'simply a metal box containing a power supply bolted to a large circuit board' that 'met the minimal definition of a computer and no more' (Freiberger & Swaine 1984).

Sold in kit form to enthusiasts, it was a runaway market success that surprised its inventors (ibid.); it 'breached the machine room door, and rivals emerged almost at once from garages all over the country' (ibid., 57).

These progressive technical developments led to major shifts in the organization of the new computer industry. In the US the mainframe industry of the 1950s and early 1960s was dominated by established firms from the accounting machine era of the third Kondratieff, notably Remington Rand (Sperry Univac) and IBM. But the minicomputer revolution of the mid-1960s, and the microcomputer revolution of the late 1970s, were both marked by the emergence of new firms which first developed and exploited the application of the new technologies: the Digital Equipment Corporation and Wang in the first, Apple, Commodore, Tandy (and Sinclair, Acorn and ACT in the UK) in the second.

This, simply, was because the established manufacturers failed to appreciate the potential of the new market. At the birth of the microcomputer both MITS and its competitors were hobbyist enterprises; none of the big companies wanted to build microcomputers:

> Without exception, the existing computer companies passed up the chance to bring computers into the home and onto the desk. The next generation of computers, the microcomputer, was created entirely by individual entrepreneurs working outside the established corporations. (Ibid., 18)

Kenneth Olsen, founder of Digital – a pioneer entrepreneur of the previous minicomputer revolution – rejected the idea of producing such a machine, saying he could see no use for a computer in the home (ibid.).

The result, inevitably, was a swarming of new companies during 1975, mainly in the San Francisco Bay area: IMSAI, Cromenco, MOS Technology, Microcomp Associates. Many, lacking entrepreneurial flair or management skills, soon crashed (ibid.). A significant element consisted of California campus dropouts from the 1960s era, who had an ideological commitment to computing for the people, and who met in the legendary Homebrew Club, which held its first meeting at Menlo Park in the heart of Silicon Valley in March 1975 (ibid.). This highly informal group – with no official membership, no dues, open to all – played a critical rôle in the exchange of ideas and the solution of technical problems at the birth of the new industry.

The outstanding commercial success story of this era was of

course Apple, founded by two campus dropouts and Homebrew members, Stephen Wozniak and Steven Jobs, to market a home computer they developed early in 1976. Built with funds acquired by selling their own possessions, the Apple I sold a mere 200 through local stores and mail order. The Apple II, which they began to develop in the autumn of that year, was eventually financed by A.C. 'Mike' Markkula, a retired millionaire marketing director of Intel; it achieved Markkula's ambition by becoming the only company in business history to pass into the Fortune 500 within five years of foundation, reaching sales of $583 million in 1982 (Freiberger & Swaine 1984, 215, Rogers & Larsen 1984).

But the spawning of new companies did not end there. For the new hardware also required specially written software to make it usable, and above all to make it usable by the millions of new customers who had no previous computer experience. Some of the resulting companies were developed, like Apple, almost literally out of nothing: Digital Research was founded by Gary Kildall, an instructor at the Naval Postgraduate School at Pacific Grove in California, after he had developed (in 1973) the first commercial operating system for microcomputers, CP/M, in advance of the appearance of the Altair; BASIC was adapted for the Altair by two Harvard students, Bill Gates and Paul Allen, who then founded Microsoft; MicroPro, which developed WordStar, was founded by a refugee from the short-lived IMSAI Corporation, the second to market a microcomputer (Freiberger & Swaine 1984).

By this time the established companies had begun to try to catch up. The first, significantly, was not a computer firm at all but a retail chain: Tandy, which was originally founded as a leathercraft company in Fort Worth, Texas, in 1927, and had moved into electronics when it acquired a small, struggling chain of radio stores, Radio Shack, in 1962. It reluctantly developed its first machine, marketing it in August 1977 and only then realizing its potential (ibid.). The colossus of the industry, IBM, announced its Personal Computer only in August 1981; it was immediately recognized in the trade as a completely conventional product, using components bought in from existing producers and running established software which had been adapted for IBM in conditions of great secrecy by its proprietors (ibid.). Imitation, evidently, was the sincerest form of flattery: IBM, a company based for 30 years on a totally different concept of making and selling computers, had to borrow all the products of its competitors.

Advances in telecommunications

Parallel to these advances in computing, and dependent in large degree on the same developments in component technologies, came major changes in the transmission of information. The invention of the transistor was in part the outcome of research efforts at Bell Laboratories to develop improved electronic devices for telecommunications systems. Since then, there have been many advances in the basic elements of the basic telecommunications system: switching equipment, transmission equipment, and peripherals. And there have been major developments in wireless communications, most notably in satellite technology.

In switching equipment, the electromechanical Strowger systems continued to be replaced by crossbar exchanges, a more efficient electromechanical system developed in Sweden in 1915, and later by the first generation of electronic exchanges developed at Bell Laboratories in 1960. The crossbar system proved more compatible than Strowger with early attempts at computerization from the late 1960s. In the UK it was little used in the public network until the late 1960s; the Post Office attempted to develop an alternative system using hard-wired solid state circuitry which could only be changed by rewiring the system and was much less flexible than fully electronic exchanges (Hills 1984). In the USA, too, the earliest electronic switching systems used wired-logic, special purpose circuitry and technologies; but these were soon being replaced by stored program control (SPC) in which computer-like processors perform the necessary switching functions (US National Research Council 1984). The first was introduced by Western Electric in the USA in 1965; since then, more than 40 SPC switches have been developed by telecommunications equipment manufacturers round the world. The design of electronic automatic exchanges was closely related to the principles involved in computers (Tucker 1978; Dummer 1983). Their use occurred most rapidly in the USA; the first digital (System X) exchange was introduced into the British network in 1981. This was a major step towards digitalization of the telephone network, allowing all types of information to be transmitted and switched in digital form.

Developments in telecommunications transmission equipment allowed cables to carry a greater number of signals at any one time, and advances in semiconductors facilitated a major shift from the analogue to the digital mode of transmission, allowing more information to be transmitted more reliably and cheaply over similar transmission lines. Digital transmission was initially introduced in urban telephone networks in the USA to relieve trunking shortages without the high cost of installing additional cables; it was later extended to medium- and long-haul trunking (US National Research Council 1984). In turn, this enhanced the possibilities of digital data handling, leading incidentally to fierce competition between computer and telecommunications equipment manufacturers in such fields as local area networks and private branch exchanges: an example of the important principle of convergence of different information technologies, originally separate, towards the close of the fourth Kondratieff.

Technological advances also brought about many changes in the nature and range of transmission media, though copper cables remained at the centre of the transmission network in the early 1980s. From the 1940s microwave was developed as an alternative to coaxial cable for long-distance transmission; by the late 1970s, it accounted for over three quarters of the annual expenditure on long-distance transmission equipment by American common carriers. The development of satellite communications, developed originally under military contracts in the 1950s and 1960s, offered an alternative to cable and terrestrial microwave systems for long-distance and international telephone traffic. Since the mid-1960s space communications costs have fallen steadily although their range and quality has increased, rendering them an increasingly attractive medium for long distance telephone traffic, as well as broadcasting and remote sensing.

The other major advance has been optical fibre transmission. It has several times the signal capacity of coaxial cable of given dimensions, greater resistance to corrosion and greater immunity to electric interference as well as being based on silica, which is cheaper and more widely available than the copper used in conventional cables. Optical cables have the potential to provide a reliable and relatively cheap basis for distributing a broad range of communications signals in addition to conventional telephony ('the wired city' infrastructure of a fifth long wave), though the economics and politics of such potential broadband systems remained uncertain in the mid-1980s.

These optical-fibre transmission systems basically involve two product groups, the fibre cables themselves and the new opto-electronic components used to transmit and receive the signals along these cables. An early lead in optical fibres was taken by Western Electric and Corning Glass in the USA, which hold many of the early basic patents. In more recent years however, a number of other large telecommunications firms, including AT&T, ITT, NEC, Northern Telecom, NTT and Philips, have developed processes that are reported to be substantially different. In the component technologies, such as light-emitting diodes (LEDs) and lasers, much of the early work took place in the USA, but Japanese firms have played a major rôle in more recent years (US National Research Council 1984).

In line with these developments, the late 1970s and the 1980s have also seen the appearance of a wide range of increasingly sophisticated peripheral devices: the increasingly automated PABXs (Private Automatic Branch Exchanges); telex, facsimile, and videotex terminals and printers; memory telephones: and modems for domestic and business uses.

The main impact of these developments is greatly to enhance the range, reliability and potential of information transmission (whatever its type) across space while greatly reducing its cost. Hence they promise large improvements in capital productivity, much like the other elements of the new information technologies (Guy 1985). The shift towards digitalization allows closer links between computer and telecommunications systems and provides the necessary infrastructure for a potential range of significant new information industries; it may be viewed as the essential condition for any new long-wave upswing (Blackburn *et al.* 1985). These technical advances also provide a basis for the development of a wide range of new information and communication services such as mobile telephone networks and value added networks.

The consumer electronics sector

At the start of the fourth Kondratieff, and for more than a decade after the end of World War II, electronics-based New IT still largely consisted in the production of consumer goods, particularly for entertainment (Wilson 1964). There was still a growing market for radios, music reproduction and, above all, television. The early 1950s saw rapid expansion of the black and white television market, based on improvements in the basic technologies developed in the mid-1930s by EMI in the UK, Telefunken in Germany and RCA in the USA and, in the UK, on the postwar resumption of public broadcasting. Production was heavily concentrated in the hands of the original innovating companies with their large R&D resources. Colour television was an innovation of this period, mainly through America's RCA, with Germany's Telefunken playing a significant rôle (Freeman *et al.* 1982, National Research Council 1984); RCA began sales in 1954 but the product was slow to sell, because black and white television was still diffusing and there were few colour broadcasts. So it was not until the 1960s that RCA's investment began to pay off – and soon after, at the end of that decade, Japanese firms rapidly won the lion's share of the American market. Their success came not merely from imitation and low labour costs but from a wide range of both product and process innovations such as integrated-circuit technology, automated techniques in assembly, testing and handling, and personnel training at all levels, backed by large-scale R&D co-ordinated by the Ministry of International Trade and Industry. These resulted in considerable advances in quality, reliability and productivity compared with their American and European competitors (Freeman *et al.* 1982, Peck & Wilson 1982, National Research Council 1984).

This pattern was replicated in a wide range of consumer electronics products such as record and tape music reproduction systems, and, perhaps most spectacularly, in the video-cassette recorder, the major innovation of the 1970s. Although this was first developed for professional use in broadcasting studios by the American Ampex Corporation in 1956, the problem remained of how to develop it as a cheap mass-produced consumer product. During the 1960s a number of leading companies (RCA and Cartravision in the USA, Philips in Europe, Sony and Matsushita and others in Japan) worked on the problem of a cheaper machine, but still they were thinking mainly of industrial and broadcasting applications. In 1970 all these reported breakthroughs in developing helical cassette machines, but it was not until 1975 that Sony marketed its Betamax and 1977 that Japan Victor (51% owned by Matsushita) launched VHS (Video Home System), which soon came to dominate. Philips was the only real European contender but its V2000 format captured only a small part of the European market, and in 1984 it announced that it would market VHS machines, thus confirming the global triumph of this system (National Research Council 1984).

Office technology in the fourth Kondratieff

The strange fact was that for the first 30 years of the fourth Kondratieff the electronic revolution had a negligible influence on information technology in the place that might have seemed most obvious: the office. Of the four major innovations that affected office work between 1950 and 1980, two, the electric typewriter and the dictation machine, were not new at all: they represented technologies developed in the preceding Kondratieff, though brought into low–cost mass production only now. The dictation machine in particular was progressively miniaturized as the result of developments in tape recording – particularly the cassette, introduced in the early 1970s – and associated electronic technology, which together permitted a massive reduction in bulk of the equipment.

Two other pieces of technology were, however, new. The first, xerography or electrostatic copying, was an invention of 1937 by the American physicist Chester H. Carlson; it was at first ignored by the major corporations (which thought it had no commercial value) and was introduced commercially only in 1959. Like the electric typewriter in its developed form, it was American and controlled effectively by a monopolist, in this case the Xerox (previously Haloid) Corporation, during its first two decades of life. A runaway commercial success, it almost immediately displaced the stencil duplicator as a method of mass copying of documents and threatened carbon paper, the traditional method of making one or a few copies. Xerography is however, an interesting case of a special kind of innovation: deriving indirectly from photographic processes, it has no very direct technical relationship to the other strands in the history of New IT.

The other innovation, the electronic calculator, was, however, very much part of the mainstream story. Introduced during the early 1960s, it was rapidly miniaturized and reduced in price, thanks to the introduction of integrated circuitry. Its first manifestation was a machine of about the same bulk as a personal computer of the mid–1980s and costing about as much (in constant prices) as a fairly sophisticated personal computer. Within less than a decade, more sophisticated hand–held models weighing a few ounces were widely on sale in filling stations and similar outlets for about the price of a take–away pizza. This was only a particularly spectacular example of the bulk and cost reductions that followed the mass production of cheap dedicated chips in the early 1970s.

Conclusions

The distinctive features of the fourth Kondratieff were thus four. Firstly, new information technology became for the first time a key carrier technology of a long wave of economic development. Secondly, it did so through the full flowering of electronics, which had been born and partially applied during the third Kondratieff but was awaiting its full potential at that long wave's close. Thirdly, it started with separate and parallel developments in basic componentry, in computing and in communications, which then became progressively united into one technological stream, particularly through the development of digital storage and transmission of information.

Fourthly, and significantly, it was earliest applied in two almost separate streams. First of all it resulted in a stream of consumer electronic products (television sets, high-fi record-playing equipment and records, tape recorders, FM radios) that followed from technological innovations of the period 1935–40, the application of which had been delayed by World War II. Then, however, the even newer advances following the discovery of the transistor were first applied to very expensive, state-of-the-art producer goods, at first mainly in military hardware, but also in early commercial computing, thence diffusing, through further technical development, and above all through drastic cheapening, to a wider range of consumer goods by the middle of the Kondratieff – a process mimicking that which occurred in the preceding long wave (the commercial application of radio) on a far vaster scale. Having completed the summary review of the development of the technologies, we turn now to their industrial impacts and in particular to the changing geography of production.

10

New IT on the world stage: the fourth Kondratieff upswing

From the start of the fourth Kondratieff, as Chapter 9 has just shown, information technology began to become one of the true carrier waves of world economic development. Just as in the previous era automobiles and electrical machinery had determined the course of high technology economic growth and the relative fortunes of the major industrial powers, so now the laurel would pass to the nation that proved most effective in bringing electronic information systems to the market place. At the outset, around 1950, the outcome of this race was by no means clear. Three at least of the then leading economic powers – the USA, the UK and Germany – had massive inherited scientific knowledge and technological competence in electronics. The most likely outcome might well have been that they shared the prize. In fact, by the end of the long fourth Kondratieff upswing, in the early 1970s, the USA was the undoubted world leader with a new competitor, Japan, coming fast up on the rails. The aim of this chapter is to tell the story of this race, and to try to account for its unexpected outcome.

It is not an easy story to tell. The problems of cross-national data availability and data comparability, that have plagued us earlier, continue to trouble us now. Particularly, rapid technological change marches ahead of the capacities of the official statisticians: new products, new industries – computers, electronic calculators – are not identified until too late. So the result will necessarily be a pastiche drawn from different sources, not a grand canvas from a single hand.

The office equipment industry

The best place to start is with office equipment, because this was an old industry, created through advances in mechanical and electromechanical technology at the start of the third Kondratieff (Ch. 6), that now began to be transformed by electronics. It consisted of four major divisions: typewriters, accounting and calculating machinery, document copying equipment, and a miscellaneous category including dictating machines, addressing and franking machines, and letter-opening machines (Hays *et al.* 1969). At the start of the fourth Kondratieff these industries remained almost exclusively electromechanical. Yet at the point where upswing turned to downswing, around 1975, the first signs were observed of technological transformation as the barriers between technologies, once considered impenetrable, began to crumble (Commission of the European Communities 1975): in photocopying the expiry of Xerox patents was already leading to the entry of new firms with new products; electronic calculators were rapidly taking over from mechanical ones; computers were being applied to word processing, and the electronic typewriter had just arrived.

Already, during the previous Kondratieff, the office machinery industry had become increasingly dominated by American firms, followed by one or two European countries such as Germany and Sweden; the UK had fallen behind. And, though British production grew rapidly in the 1950s and 1960s, relatively the country continued to lag. Its share of total output from the 15 leading capitalist economies fell significantly between 1954 and 1966 – indeed more sharply so than its share of total manufacturing production. In typewriters its share fell from about 9% to 3%, taking it from third place behind the USA and West Germany, to sixth place behind Italy, the Netherlands and Canada. Its share of 'other' office machinery fell from 10% in 1954 (when the USA had about two-thirds of total world production) to 4% in 1966: about level with Italy and Japan, but behind Germany and France (Table 10.1).

This is a typically internationalized high-technology industry, with large trade flows between the advanced industrial nations. British exports of office machinery (excluding copiers) increased substantially between the early 1950s and late 1960s (Table 10.2): about 32% of output was exported in 1948, 54% in 1968. The share rose rapidly in the early 1950s, partly due to government exhortation, partly to allocation of materials and minimum export requirements for foreign firms established in Scotland after the war;

Table 10.1 Shares of main producing countries in output of 'other' office machinery, 1954–66[a]

	Percentage share			Value, 1966
	1954	1960	1966	($m)
Canada	4	3	2	(75.0)
USA	67	66	70	4156.4
Japan	1	2	4	250.9
France	5	6	5	322.2
Germany	7	6	8	448.9
Italy	3	4	4	(224.5)
Sweden	3	2	2	111.4
UK	10	8	4	222.3
others[b]	1	2	1	58.7
total	100	100	100	

Source: Hays *et al.* (1969).

Notes:

[a] Excludes typewriters, dictating machines, offset litho and office document copiers; includes electronic computers.

[b] Austria, Denmark, Finland, Netherlands, Norway and Spain.

The figures in brackets are estimates.

In 1966 the value of typewriter production in the 10 leading countries amounted to about $680 million, about one-ninth of the value of other office machinery as listed above.

Table 10.2 Exports and imports of office machinery, UK, 1954 and 1968.

	1954 (£m)	1960 (£m)	1968 (£m)
Exports			
computers[a]	—	1.1	43.6
other machinery[b]	12.2	27.0	47.7
total	12.2	28.1	91.3
Imports			
computers[a]	—	n.a.	51.9
other machinery[b]	6.4	21.0	73.3
total	6.4	21.0	125.2
Net trade balance[c]			
computers	—	n.a.	−8.3
other machinery	+5.8	+6.0	−25.6
total	+5.8	n.a.	−33.9

Source: Hays *et al.* (1969).

Notes:

[a] Includes peripheral equipment for computers.

[b] Includes typewriters; accounting, calculating, adding etc. machines; cash registers; dictating, addressing, statistical machines.

[c] Exports versus imports; import and export data may not be entirely comparable.

it remained fairly stable from 1954 to 1962, then again rose. In 1968, total exports of office machinery were £91 million, including £15 million for copiers and £43 million for computers.

Despite a growing absolute level of exports and share of exports in national output throughout the 1950s and 1960s, the UK's share in the OECD countries' exports of office machinery over this period (including computers, but excluding document copiers) declined from about 14% in 1954 to 11% in 1966. In 1954 the UK's share of OECD exports was comparable to that of Germany, both countries sharing second place behind a clear lead by the USA. By the mid-1960s, the USA's share had fallen from about 40% to 30%, Germany had increased its share, but the UK had slipped down to the level of France and Italy (Hays *et al.* 1969). This comparison is based on data that includes electronic computers, in which at the time the UK was performing relatively well.

Accompanying this falling share was a dramatic increase in the import of such equipment into the UK. The share of imports in total consumption of office machinery (excluding computers) increased from about 23% in 1954 to almost 66% in 1968 (ibid.); this increased penetration affected almost all sub-sectors of the mechanical and electromechanical office machinery industry. A survey of some major office machinery users in the late 1960s, which sought to identify the possible reasons for purchasing imported goods, instanced the absence of home-produced equivalent machinery, the greater efficiency of imported products, and the lower prices of imports.

For this weak performance on the international stage there are a number of explanations. A critical one was that, in comparison with other major industrial countries, the UK was slow and ineffectual in mechanizing office work. This process did not start anywhere until the turn of the century, first with typewriters, followed by calculators in the 1920s, more sophisticated accounting and statistical machines in the 1930s, and document copying and dictating machines after World War II (ibid.). But the rate of diffusion in the UK was slow, in comparison not merely with the USA but also with major European countries (ibid.). Not only did most of the new technologies originate in the USA; American companies started to apply them commercially and on a large scale, even when they originated elsewhere.

European countries such as Germany, Sweden, and France were relatively quick to follow the American lead, even though, as in the Swedish case, their home market might be much smaller than the British. Even in the 1920s and 1930s it was estimated that the average German firm was better mechanized than its British

Table 10.3 Estimated sources of office machinery supply in the UK, 1964–6.

Type of machine	Percentage of total by value (responding firms) British-owned firms	Foreign-owned firms	Imported
typewriters	37.6	33.8	28.6
accounting machines	2.0	89.7	8.2
adding and calculating machines	8.5	61.9	29.5
cash registers	5.3	72.6	22.1
dictating machines	2.3	39.2	58.4
document copiers	23.0	19.1	57.8
duplicators	69.3	15.5	15.2
statistical and other	39.8	31.8	28.4

Source: Hays *et al.* (1969).

counterpart. In the mid-1960s, usage of office machinery per head of the labour force was three times as high in the USA as in the UK, in Sweden twice as high, in Germany two-thirds greater and in France one-third greater. The ratio of typewriters to population was estimated as 1 to 5 in the USA, 1 to 30 in continental Europe and 1 to 55 in the UK. One of the main reasons was that, at least until 1939, labour for office work was relatively abundant and cheap (ibid.), an explanation that has frequently been raised for the poor technological 20th-century performance of British industry generally (Kirby 1981). It suggests a very important conclusion: that a high rate of *process* innovation among home consumers may spur a high rate of *product* innovation among producers, which in turn may produce a base for exports.

One outcome is that even British domestic production has been dominated by overseas-based, especially American, firms (Table 10.3). These seem to form the spearhead in most aspects of technical development, and their productivity is considerably higher on average than that of British-owned firms (Hays *et al.* 1969, Commission of the European Communities 1975). By the late 1960s, foreign-owned firms accounted for about 60% of the employment and about 75% of output of the industry (Hays *et al.* 1969). In 1972, 7 of the 9 largest office machinery manufacturers in the UK were overseas-based companies; 12 of the top 20 enterprises were subsidiaries of foreign firms (Commission of the European Communities 1975).

The origins of this dominance lay in the initial introduction of most of the major new products. In most cases this occurred in the USA or other foreign countries, so that initial production in the UK was by the subsidiary of an overseas firm. This was early true of typewriters, where American and German firms dominated (Clapham 1938, Hays *et al.* 1969, Prais 1981). Business and accounting machines, similarly, were largely produced in the UK by American firms; thus cash registers were first produced in the USA by the National Cash Register Company, which set up a subsidiary in the UK as early as 1895 and started production on a substantial scale in the UK at Dundee in 1946; Burroughs set up factories at Strathleven in 1951 and Cumbernauld in 1958, and Remington Rand established a factory in Glasgow. The Dictaphone Company began business in the UK in 1947 in order to import dictating machines made by the American firm of Dixon Brothers, and later assembly and complete production facilities were established at Acton, moving thence to East Kilbride. It remained wholly American owned, and by the late 1960s there was no British owned or controlled production of dictation equipment (Hays *et al.* 1969).

In mechanical and electromechanical punched card machines, production in the UK between the wars was undertaken by two companies: the British Tabulating Machine Company, manufacturing under licence from IBM, and Powers-Samas, under licence from Remington Rand. After the war the agreements were terminated because of pressure from the British government to save dollars (and avoid American Anti-Trust legislation); at last, a number of British firms entered this field, which gradually merged into the growing computer industry. In addressing machinery, although a relatively unsophisticated machine was produced in the UK as early as 1900, the field was dominated by American and other foreign firms after the invention of the Addressograph machine in the USA in 1930.

The major exception seems to be duplicating machinery, although this had ceased to be a leading technology by the 1960s. The rotary stencil machine, the oldest type of duplicating equipment, appears to have been a British invention, and the three leading British firms in the 1960s, Gestetner, Roneo and Ellams, began production towards the end of the 19th century (ibid.). British firms continued to be involved in the production of electronic stencil-cutting devices, and the spirit duplicating field was shared by indigenous and foreign firms. The use of offset litho machines for office duplicating purposes began with German-built machines, but during World War II indigenous firms began production with government support; afterwards, production was shared between British firms working under American licences and foreign-owned firms (ibid.). In the field of electrostatic copying equipment, which grew rapidly in the 1960s, the major British producer was Rank-Xerox, owned jointly by the Xerox company of the USA and the British Rank Organization.

There was one postwar British innovation: in 1961 a British company, Sumlock-Comptometer, produced and marketed one of the first electronic desk calculators, the Anita. For several years it was estimated to command half the British market. The firm was taken over by another British firm, Lamson Industries, later in the 1960s.

There are a number of reasons for this foreign dominance (ibid.). Firstly, as already seen, American firms pioneered the technology and brought it to the UK. Secondly, in the 1960s, foreign-owned production was extended to some of the most mature and established products such as typewriters, with overseas firms buying out British concerns or setting up concessionaires to import their products. Thirdly, in the more sophisticated products, foreign firms were sophisticated technologically, with higher R&D levels than their British equivalents. Fourthly, regional policy encouraged the establishment of foreign firms, especially in Scotland. Fifthly, the single-product nature of most British firms, combined with the economic advantages of diversified production and product ranges, provided an additional incentive for the absorption of indigenous firms by larger, more diversified, international organizations. Sixthly, the structure of demand for many of these products is such that direct selling by foreign manufacturers is not practicable; the establishment of local subsidiaries, agencies or concessionaires becomes a preferred option. And seventhly, foreign firms had good sales organizations in the UK, sensitive to user needs in the development of new products, and lending equipment on a trial basis (ibid.).

Underlying the competitive power of the foreign companies, however, were also some basic and only too familiar facts. They tended to be large, well integrated companies offering a wide range of products, with a good and extensive servicing back-up; so that once they had sold their systems to major buyers they could rely on their continuing loyalty. In the face of competition from such established giants, entry by newcomers would indeed have been difficult. They had more capital equipment per employee, and consequently higher productivity. All this was true whether they were operating factories in the UK, or in their home country. In fact, had they not established British subsidiaries, the UK's export performance would probably have been worse, and its import penetration problem greater, than it actually was in the 1960s. And, when some of the multinationals began to transfer operations to other European countries at this time, the fact began to show in the performance of the British office equipment industry as a whole (ibid.).

Electronics capital goods

The early growth of the electronics industry was based on consumer goods such as radio and television, although the interwar era did see the emergence of electronics capital goods in the form of radio transmitting equipment (both civil and military), as well as some application of electronics in telephone communications systems and in new fields such as radar. But the real growth of electronics capital goods came only after 1950. Between the mid-1930s and the mid-1960s electronics production grew worldwide at over 10% per year (in constant prices). But, from the late 1950s, the growth rate for consumer goods (primarily radio and television receiving sets and music recording and reproducing sets) began to slip behind that of capital goods, which was achieving approximately 15% per annum (Wilson 1964, NEDO 1967, 1968, 1970, Freeman *et al.* 1965).

It is not always easy to define or to measure this group of industries, because of the rapid pace of technological change and the fact that capital and consumer goods use many of the same components and are even produced in the same plants (NEDO 1968, 1969, Freeman *et al.* 1965). But the following classification, based on Ministry of Aviation and other official sources of the 1960s, is a useful guide:

(1) electronic computers (or data processing equipment): digital and analogue computers and their peripheral equipment.
(2) radar and other electronic navigational aids: ground, marine and airborne radar systems and other electronic navigational aids and guidance systems, missile and satellite tracking and detection equipment and sonar.
(3) radio communications and public broadcasting and ancillary equipment: radio communications equipment for ships, aircraft, police, ambulances and services equipment; transmitters, aerials, audio and video studio recording equipment, monitoring and studio equipment, and relay and repeater systems for radio and television broadcasting systems.
(4) electronic and nucleonic measuring and testing equipment: electronic testing and measuring including oscilloscopes, signal generators, spectrometers, radiation detectors, computer

and component testing equipment, sound and vibration measuring equipment.

(5) industrial electronic control equipment: machine-tool control equipment, other electronic control equipment including detectors of foreign bodies, crack and strain detectors, counting, weighing and sorting equipment. This subsector produces key elements of the 'tertiary' or 'control' mechanization equipment which only begins to experience a rapid rate of diffusion from about 1960 (Blackburn *et al.* 1985).

(6) other electronic capital goods: a wide range including electronic welding and heating equipment, electronic medical equipment, electronic simulators and training aids, electron microscopes and ultrasonic equipment.

(7) telecommunications equipment. Although much of this equipment was predominantly electromechanical rather than electronic throughout this period, some studies include it within electronics capital goods (Freeman *et al.* 1965). It includes telephone and telegraph apparatus, telephone sets, switchboards, teleprinters, telewriters, telegraph and telephone testing equipment.

Though all these divisions grew during the fourth Kondratieff upswing, there were significant differences in rate, timing and geographical distribution. Here we shall consider merely some representative examples.

Radar and associated products

These have been closely affected by military demand, though air traffic control created a major civilian market also (Freeman *et al.* 1982). By the mid-1960s, this was the largest single category of electronics capital goods production in both the UK and the USA. Although the UK appeared to have a slight technological lead down to 1940, with Germany and France also having strong technological capacity, by 1950 the USA commanded a clear lead in levels of production and shares of world exports, with the UK in second place; West Germany and Japan had negligible shares because of postwar limitations on their defence industries.

American dominance reflected the strength of domestic military demand, which according to one estimate was no less than 90% of the total market of all the advanced capitalist economies in the mid-1960s; on this basis, it also commanded more than half of total world exports in the late 1950s and early 1960s (Freeman *et al.* 1965). By the mid-1960s the UK produced about 5% of the advanced capitalist economies' total output, but commanded about 29% of total exports. Some British firms such as Marconi and Decca were particularly successful in marine and airborne radar and navigation systems in this period, exporting a very high proportion of their output. It was estimated that outside the American continent about three-quarters of the ships in the world equipped with marine radar had British systems and that most of the civilian airports were similarly equipped (ibid.). British firms were in fact so strong in this field that they were able to license French, German and other electronics firms.

Computers

Even in the mid-1960s, despite extremely rapid rates of growth since the mid-1950s and especially since the emergence of 'second generation' machines incorporating transistors from about 1958, computers still stood second to radar in importance within this category (Harman 1971, Margerison 1978). The origins of the industry, in the late 1940s and early 1950s, lay in research and development work in universities, government and industrial research laboratories, especially in the USA and the UK, to a lesser extent in Germany, France, the Netherlands and Italy. Much of this work was sustained by the enthusiasm of groups of scientists and engineers and by sponsorship from a variety of government agencies.

The general commercial view in the USA, at least down to 1950, was that there was little or no market. Thomas J. Watson Senior, the head of IBM, despite his long acquaintance with business needs for data processing, felt that a few machines could deal with the total needs of those requiring complex calculations; nor was there any broad demand among major life assurance companies, telecommunications firms, aircraft manufacturers and others who were informed about the emerging technologies (Katz & Phillips 1982, Freeman *et al.* 1982). For IBM's 650 computer, sometimes described as the Model T of the industry, the engineers forecast a sale of 200, whereas the firm's product planning department forecast no sales at all. After its introduction in November 1954, 1800 machines were sold (Harman 1971).

Thus much of the enormous progress in computing technologies in these early postwar years came through military and other government support, including grants to universities; industrial support played a relatively minor rôle. Among those who first saw the commercial potential of computers were the pioneers Eckert and Mauchly, who made great efforts to interest financiers and private companies after their dismissal from the University of Pennsylvania in 1946 because of their commercial activities. They experienced serious difficulties; Remington Rand, which acquired their firm in 1950, tried to cancel all outstanding contracts because it took such a pessimistic view. Yet from this time it became clear that there was a commercial potential for computers. In the USA the Korean war increased military and other government support, leading to IBM's keen interest and full involvement. IBM now soon demonstrated that the computer industry could be extremely profitable; it rapidly became one of the most profitable and fastest-growing companies in the world, accounting for between 60 and 70% of world computer sales throughout the period 1950–80. Only in minicomputers and specialized systems was its share of the world market significantly below 60%. By the mid 1960s, only in the UK and Japan among the larger capitalist economies did its share of installations fall below half (OECD 1969, Harman 1971, Stoneman 1976, Freeman *et al.* 1965, 1982).

During this period there was limited swarming as a number of new as well as established electronic firms entered the industry; by 1960 there were about 26 firms in the USA, 7 in the UK, 3 in Germany, 2 in Holland, one each in France and Holland engaged in the design, construction and sale of computers (Margerison 1978). An international survey around 1970 suggests that there were about 39 firms active in the computer industry in the USA; among the other major countries there were 7 significant domestic computer firms in Japan, 4 in Germany, one each in the UK, France, Italy, the Netherlands, Sweden, Norway, Denmark, Israel. Many of these 'home producers' had licensing ties with the larger American firms (Harman 1971). Many other firms entered the industry in the 1950s but had bowed out by the late 1960s. Thus, though limited swarming occurred in the 1950s and 1960s, not much of it proved profitable.

Though several European governments did try to keep national producers afloat, by the late 1960s most had been forced to withdraw or amalgamate. By then the only significant non-American producers designing and developing their own equipment in Europe (other than the UK) were making small scientific computers or specialized systems. In Japan the share of American-owned firms was lower than in Europe, largely as a result of state policies (OECD 1969, Harman 1971, Freeman *et al.* 1965). Thus the computer industry in this period was an extreme example of Schumpeter's Mark 2 model (Freeman *et al.* 1982).

The clear American leadership during the 1950s and 1960s stemmed from the size of the domestic market. By 1965 the USA had an estimated 20 000 computers installed; Japan had 1840, West Germany 1603, the UK 1160 and France 1061. Thus the USA had 105 computers per million of population in 1965, Western Germany had 29, France 22, the UK 21 and the Netherlands and Japan 19 (Freeman *et al.* 1965). As earlier in mechanical and electromechanical information, the USA took a clear lead in mechanizing or automating information handling. Although there were no satisfactory comparative statistics for production and trade in computers even by the mid-1970s, one survey suggested that American domestic production amounted to about US $1 billion by 1964, representing about 71% of the total combined production of computers in the five leading countries, the USA, the UK, West Germany, France and Japan (ibid.). The American industry's exports of electronic data-processing equipment amounted to $218 million in 1964: almost three-quarters of the combined exports of the five leading countries. Its dominance in this subsector was more pronounced than that in the electronics capital goods sector as a whole.

Radio communications and public broadcasting etc. goods

During the late 1950s and 1960s, this subsector represented almost as large a market as electronic computers. And here again the USA dominated, with about three-quarters of world production and almost 60% of world exports in the mid-1960s. Again the data on world production shares and exports substantially understate the rôle of American companies. International Telephone and Telegraph was one of the leading manufacturers and exporters in this field as well as in telephone equipment; the company employed 185 000 people in 1964, 128 000 of them in Europe where subsidiaries included Standard Telephone and Cables (STC) in the UK and Standard Elektrik Lorenz (SEL) in Germany (ibid.).

British production in this field was about one seventh that of the USA in 1958, falling back to about one-twelfth by 1964; by then its percentage share of total production of the five leading countries was 6.4%, slightly ahead of West Germany and France, and its share of world exports was somewhat higher at 14.2%. Racal was one specialist British firm which enjoyed very rapid growth and export markets in this field, as did Plessey and Marconi. Other European firms with a strong export performance in this field included Philips, Telefunken, Siemens, CFTH, Pye and Redifon (ibid.). From the late 1950s Japanese firms experienced fairly rapid growth of production and exports in this field.

The public broadcasting and studio equipment element in this category was relatively small compared to the specialist radio communications market. Here the USA accounted for about 60% of world output and about 40% of world exports by the mid 1960s. As in the other category of radio communications equipment, European firms had fairly strong technology and patent positions, with a considerable degree of two-way licensing with American firms. The major European firms, such as Telefunken, Marconi, Siemens and CFTH, all enjoyed a strong position in world export markets. This indeed was the one area of electronic capital goods where there are significant exports from Europe to the American market (ibid.).

Summary

By the mid-1960s the clear picture was one of absolute American dominance, albeit with some variation between subsectors (Table 10.4). This dominance is all the greater if account is taken of the contribution of American companies in other countries through subsidiaries, technology licensing agreements and the like.

Table 10.4 shows that in the mid-1960s the American share of total electronic capital goods production in the five leading countries was almost 82%; its share in each of the subsectors ranged from 91% in radar and navigational aids to a 'low' of 61% in

Table 10.4 Production of electronic capital goods, national shares, 1964.

	USA	UK	West Germany	France	Japan
Electronic data-processing equipment	70.8	9.0	7.4	7.7	5.1
radar and navigational aids	90.9	4.4	0.5	3.4	0.8
radio communication and broadcasting etc.[a]	76.5	6.4	6.3	6.0	4.9
electronic testing and measuring[b]	73.5	8.2	8.2	7.4	2.7
industrial electronic control	61.3	12.6	13.8	11.1	1.2
other electronic equipment not elsewhere classified	82.3	4.1	5.1	4.1	4.4
total electronic capital goods	81.9	6.0	4.2	5.1	2.8

Source: National Institute Economic Review, **34**, 1965, 42.

Notes:

[a] Includes radio communications equipment, public broadcasting, transmitting and ancillary equipment.

[b] Electronic measuring, testing and analysing instruments, including nuclear instruments and controls.

industrial electronic control equipment production. The UK's share of total production in this sector was second highest, but it was a long way behind at 6%. Within the different subcategories, the UK's highest share was 12.6% in industrial electronic control equipment followed by 9.0% in electronic data processing equipment. France was in third position with 5.1% of the total; West Germany was fourth with 4.2%, Japan fifth with only 2.8%. By this time Japan's capital goods and consumer electronics industries were beginning to grow rapidly; their relative international position would improve markedly over the following years, as we will see in Chapter 11.

In terms of shares of total electronics capital goods exports, the USA again commanded a leading position with about 60% of the total. It commanded a 75% share of world exports of electronic data-processing equipment and 70% of the other electronic equipment category; its lowest share of the five countries' total exports was in industrial electronic control equipment with 29%. The UK was again in second position behind the USA, commanding 14.5% of the total exports of the five countries, a much higher share than of total production. Among the subcategories, its largest export share was in industrial electronic control equipment, where it had 35.8% of the five countries' combined exports, and in radar and navigational aids, where it had 29.1%. Western Germany was in third place, with 12.7% of the total; France was in fourth position with 7.3%, Japan in fifth with 5.0% (ibid.).

Telecommunications equipment

Throughout the fourth Kondratieff upswing telecommunications remained a far from fully electronic technology. Yet it was one of the basic information technology industries. Even then, there was a two-way flow of technological innovation between it and electronics; a process that began to accelerate in the mid-1960s.

The production of telecommunications equipment, like that of so much other information technology, was overwhelmingly dominated by the USA (Table 10.5). But, though American production far exceeded that of the next four countries and indeed amounts to approximately two-thirds of the five countries' total production, it was primarily for domestic markets. Indeed, British exports well exceeded those of the USA in both 1958 and 1964. This was not due to any technological or organizational weakness on the American part; it is explained by barriers to trade created by the traditional structures of national telecommunications service systems, and by related political and national security considerations.

Table 10.5 Telecommunications equipment: estimated production and exports, major countries, 1958–64.

	USA ($m)	UK ($m)	West Germany ($m)	France ($m)	Japan ($m)	Total ($m)
Production						
1958	1297	146	171	90	70	1779
1964	1956	270	386	218	252	3091
Exports						
1958	46.0	65.9	35.9	8.9	7.3	164.0
1964	38.8	92.5	91.8	17.1	13.6	253.8

Source: National Institute Economic Review, **34**, 1965, 44.

In almost every developed country, telecommunications services have historically been organized as a monopoly, usually state-run, and equipment sales have also been highly concentrated. Western Electric, the largest American manufacturer and part of AT&T until the 1984 breakup, focused almost exclusively on domestic sales to Bell System operating companies. Therefore, although its innovations have offered continuing improvements to telephone services in the USA, there has been little export spinoff.

Thus, trade figures tend to understate the performance of American firms and technologies in this industry since the war (Freeman *et al.* 1965, US National Research Council 1984). Particularly, they fail to show that American firms have enjoyed success in foreign markets through their subsidiaries. For example, International Telephone and Telegraph (ITT) has historically enjoyed a strong position in Europe. In 1925 ITT acquired International Western Electric, which held a strong patent position in European telephony and had developed close ties with service provider organizations. ITT capitalized on these ties and subsequently attempted to develop similar ties in less developed countries. Although they have in some cases shared a common core technology, ITT subsidiaries have often presented themselves as relatively autonomous (US National Research Council 1984).

Nevertheless, it remains significant that the telecommunications industries in the UK and West Germany exported considerable shares of their total output and commanded rather high shares of the total exports of the leading producing countries, with the Japanese industry beginning to expand fairly rapidly from the mid-1960s. In the 1950s the major British producers were GEC, AEI and Marconi (the latter two later taken over by GEC), Plessey, Thorn and two subsidiaries of foreign firms, STC and Ericsson. By the 1970s the major indigenous producers were GEC, Plessey, Thorn-GTE (a joint USA–UK venture); the major foreign-owned firms were STC, PYE-TMC, Ericsson, and IBM which supplied a major share of the private exchange market from the mid-1970s (Hills 1984).

Electronic components and semiconductors

As seen in Chapter 9, it took over a decade for the transistor to make a major impact on electronics outside military applications; for much of the fourth Kondratieff upswing, more traditional electronic valves and similar components, involving relatively labour-intensive processes, constituted the major part of the electronic components industry (Wilson 1964). Yet the emerging semiconductor sector of the industry represented a key field for the development of electronics, and thus of New IT, at this time (OECD 1968).

Before World War II, European researchers kept abreast of their American counterparts in electronic component innovation. Some of the most important advances in valves and tubes were made by firms such as Philips, Telefunken, GEC, AEI (BTH) and Marconi, and by European universities. There was a strong two-way flow of licensing and patenting agreements across the Atlantic (Maclaurin 1949, Sturmey 1958, Freeman *et al.* 1965). Yet by the postwar boom era this position had changed completely; almost all the important inventions and innovations were being made in the USA. This was not simply because the crucial inventions – the transistor, the integrated circuit, the computer-on-a-chip – took place in the USA. Invention was only the first step in the commercial production of smaller, more reliable, compact and energy-efficient components; successful mass production needed massive research and development resources and the accumulation of a wide range of minor process and product innovations through what Rosenberg (1982) terms 'learning by doing and learning by using'.

As with computers, governmental military and space programmes played a major rôle at this critical stage. Throughout the 1950s about two-thirds of the total R&D work of the large American electronics and aircraft firms was state financed (Freeman *et al.* 1965). By 1955 the American government was buying 22% of all semiconductor production and accounted for 35% of the value of the semiconductor market (Braun & Macdonald 1982). Indeed, the government purchased almost all the 90 000 transistors produced in the USA in 1952. It has been estimated that in 1955 total state

R&D funding to the American semiconductor industry amounted to $3.2 million whereas production refinement funding for transistors and diodes totalled about $4.9 million (ibid.). Defence requirements accounted for about a third of the value of all semiconductor sales in the USA in the mid-1950s and grew to about half by the late 1950s; the share decreased to 30% by 1966, to 24% by 1972 and to about 10% later in the 1970s (ibid.)

Initially, the established electronics firms may have benefited most from these expenditures, but by the late 1950s a large share of the military market went to the newer semiconductor firms, a fact that may have been essential for their growth (ibid.). The military market was particularly valuable because of its enthusiasm for the latest and best components at almost any price. This meant that new, expensive products could be tested and sold. Some had no civilian applications; many others had, and components engineered to meet stringent military requirements could be adapted for less stringent civilian ones. There can be little doubt, then, that military R&D support and procurement contracts helped firms to gain production experience and lower costs, thus speeding up the 'learning curve' among American semiconductor manufacturers (Freeman *et al.* 1965, OECD 1968, Tilton 1971, Braun & Macdonald 1982).

But, of course, military expenditures are only part of the story. Large industrial R&D resources, together with easy access to Bell Laboratories patents in return for a relatively small fee and the availability of venture capital for new firms led by personnel who had spun off from established firms, all played a part in the phenomenal advance of the technology and growth of the industry in the USA. The American semiconductor industry since the war is often cited as a classic demonstration of the importance of new, initially small, firms using new technologies and managerial techniques, overcoming the technical and managerial conservatism of existing firms through radical innovations and building new industries: the Schumpeter Mark I model of swarming and growth. Certainly, compared to the mainframe computer industry, new firms have played a greater rôle in the American electronic components industry. There were a large number of new entrants between the mid-1950s and the late 1960s, accompanied by major changes in firm shares of domestic and international production. But the mushrooming process may have ceased in the USA by the late 1960s: it is estimated that there were about 15% fewer firms in the industry in 1980 compared to a decade before (Braun & Macdonald 1982). It is also important to note that the large established American electronics firms with large R&D resources were responsible for about half the major innovations from 1950 to 1980 and for more than half the patented inventions and process innovations (Freeman *et al.* 1982).

There was no similar swarming of new firms in the semiconductor industries of Europe or Japan. In the UK and other European countries very few new small firms entered the semiconductor industry; the international diffusion of the new technologies came largely through subsidiaries of American firms or through large established indigenous electronics firms (OECD 1968, Tilton 1971, Sciberras 1977, Braun & Macdonald 1982). Down to the late 1960s, the American semiconductor industry was about the only international one; in other nations, firms were largely concerned with retaining a share of their own domestic markets (Braun & Macdonald 1982). It was estimated that, in 1970, American firms had 56.5% of the world semiconductor market, European firms 16.1% and Japanese firms 27.1% (Braun & Macdonald 1982).

After a relatively slow start compared to the USA, semiconductor consumption in Britain, West Germany, France and Japan rose quite quickly from about 1960. Simultaneously, between the late 1950s and the early 1970s, American semiconductor firms launched a major programme of establishing their own factories in countries such as the UK, France, West Germany and to a lesser extent Japan. This brought a decline in the share of consumption supplied by imports from the USA. Thus between 1968 and 1972 direct exports from the US to these four countries increased by 102% and the value of the produce of American assembly and fabrication plants rose by about 360% (ibid.). The percentage share of their combined semiconductor consumption that came from American imports rose from 11% in 1960 to 32% in 1967 and fell to 18% in 1972. The cumulative number of American-owned semiconductor fabricating and assembly plants in these four countries rose from 9 in 1960 to 18 in 1967 to 52 in 1972 (ibid., see also Tilton 1971). Additionally, many indigenous firms manufactured under licences from American firms like Texas Instruments, Fairchild and Western Electric (Bell Laboratories). With the exception of Philips, European semiconductor firms were largely excluded from competing in the markets for mass-produced semiconductor components and were largely confined to their national markets or to the production of specialized and customized product for niche markets (Sciberras 1977).

Compared to Europe, the major mechanism of semiconductor technology transfer from the USA to Japan in the 1960s and early 1970s was licensing agreements between Japanese and American firms. Until 1974 Japanese government policies were strongly opposed to the establishment of subsidiaries by foreign-owned firms, and overseas companies were generally not allowed to take controlling interests in Japanese firms. The Japanese government controlled the terms of the licensing agreements between Japanese and foreign firms, promoted R&D programmes and spread the findings throughout selected parts of the industry, seeking thus to maintain a balance between the benefits of large oligopolistic firm structures and those of competition (Tilton 1971, Braun & Macdonald 1982). For these and other reasons, Japan's semiconductor production, consumption and exports grew quite rapidly throughout the 1960s and 1970s, mostly through indigenous firms. In 1977, for example, although firms based in the USA accounted for 51% of European production of discrete semiconductors and integrated circuits, they accounted for only 10.7% of Japanese production. The 1970s witnessed a significant growth of the Japanese semiconductor industry, still largely using American technology; but in more recent years Japan has begun to establish a lead in the production of certain types of semiconductors, thus causing much concern to US firms and policy makers (US National Research Council 1984).

The late 1960s also witnessed a major expansion of overseas investments by American (and later other) semiconductor firms to less developed countries, especially the so-called Newly Industrializing Countries (NICs) of Asia and Latin America. The major reason was the search for cheap labour for what was still a relatively labour-intensive industry; the light weight per unit value and relatively cheap transport costs of these components also facilitated this new international division of labour. Most of such plants were confined to the particularly labour-intensive assembly operations; the more complex and technology-intensive fabrication processes remained in the developed countries.

Curiously, since the mid-1960s there has also been a reverse investment flow, from Europe and Japan into the American industry (Braun & Macdonald 1982, OECD 1985). Part of the reason has been to acquire a more direct line of access to the latest technologies and to tap into the more intangible but nevertheless informal networks of information that have played a major rôle in the dynamism of the American industry (Rogers & Larsen 1984). Another has been to facilitate access to what remains the largest single semiconductor market in the world (OECD 1968).

It is difficult to generalize about the international structure and geography of the industry at the end of the fourth Kondratieff upswing (ibid.). Within the more traditional product lines the major producers showed much greater equality in terms of

Table 10.6 Electronic component production, major countries, 1965.

	Tubes ($m)	Semiconductors ($m)	Passive components ($m)	Total ($m)
USA	1019	1012	1259	3290
Japan	137	140	390	667
UK	n.a.	208	454	(662)
Germany	122	47	404	573
France	96	59	264	419
Italy	n.a.	25	n.a.	(92)

Source: OECD (1968).

Note: Because of great differences in national definitions, these data should be regarded as rough estimates only. Data for passive components are probably too high except for the USA.

indigenous technological capacity, relative levels of production, and trade performance than in the growing semiconductor subsector. In receiving tubes, the Netherlands and West Germany enjoyed an international leadership position for some products, whereas Japan appeared to have a strong position in certain kinds of passive components. But in the dynamic area of semiconductor technologies and associated industrial activities and firms, the USA commanded a clear lead (Table 10.6). The 'technological gap' in innovation of a particular technology in different industrialized countries may have narrowed over the 1960s (ibid., Braun & Macdonald 1982); yet enormous differences remained between the American semiconductor industry and those in other advanced industrial economies. The narrowing of the gap, in large measure, occurred through the transfer of American technology via either direct exports, licensing agreements or direct investment. By the end of the 1960s, the Japanese industry had begun to advance particularly rapidly; and there was already a significant relocation of investment and employment into a select number of less developed countries.

Conclusion

Overall, then, the fourth Kondratieff upswing saw a significant shift in the international division of labour within the information technology industries. Very rapidly, the USA achieved undisputed world hegemony; Europe in general, the UK in particular, found its earlier position eroded. In the older mechanical and electromechanical office equipment industries, the UK found itself increasingly displaced, above all by the USA, partly also by its European neighbours. In electronic capital goods, too, the USA had a clear lead both technologically and entrepreneurially. The UK maintained a position in computers, but was still far behind the USA. In electronic components British producers found themselves reduced to specialist niches; the Americans, again, were clearly in the lead. Japan was just emerging as a presence in several of these lines; some areas, such as component assembly, were being located in NICs. At the end of the upswing, just after 1970, the stage was already set for the next act of the drama.

11

Information technology in the world economy: the fourth Kondratieff downswing

Some time in the early 1970s, triggered by the oil supply crisis of 1973–4, the long fourth Kondratieff boom gave way to the downswing. In the advanced industrial economies generally, employment growth in manufacturing – which had been slowing markedly during the later upswing years – was succeeded by stagnation, even by decline. The impact of this transition varied from one national economy to another, depending on their structural strengths or weaknesses: in the UK, where factory employment was already contracting, it shrank faster; in Japan, which still enjoyed vigorous growth, it slowed. But everywhere the change was evident.

It raises a critical question, central to this study. In such a transition, do the advanced-technology industries contradict the prevailing trend? Do they perform an historic function as the carrier wave, the precursor of the next Kondratieff upswing? Most studies of long waves, oddly, avoid this question. The major exception, the work of Freeman *et al.* (1982), suggests a contrary hypothesis: new technologies contribute new jobs, through product innovation, only in Kondratieff upswings; in the downswings, by aiding process innovation, they actually displace jobs and contribute to rising unemployment, even while their output continues to grow. In support, the Freeman group cite empirical data for the USA, the UK and Germany (ibid.).

Their argument seems to contradict the evidence from previous long waves, especially the third, when the then advanced-technology industries (radio, early electronics) displayed vigorous growth through the recession and depression phases (Ch. 6). It also disagrees with a considerable recent literature, which suggests that the new information technologies have special qualities: 'pervasiveness' (widespread applications in production of goods and services), 'capital saving' (increased capital productivity) and thus 'profitability' on invested capital (Mandel 1975, Soete & Dosi 1983, Blackburn *et al.* 1985, Perez 1985, Soete 1985a, 1985b). Thus, although the life cycle of the electronics-based information-technology industries precisely corresponds to the fourth Kondratieff – and, according to the classic model, started with innovations in the depression phase of the previous long wave – these industries are now experiencing an extension of that cycle, as in turn they now combine to create the technological basis of the coming fifth Kondratieff.

More precisely, even though the new technologies are displacing jobs in the sectors to which they are applied – in factories through control processes and automation, in offices through word processing and data systems – their very production will create some new jobs in the advanced-technology sectors. This is particularly so, since their major applications are coming precisely in those service sectors where capital intensity and hence productivity have until recently been low, as evidenced by continued growth in demand for information technology in offices. These have been the sectors continuing to record growth in employment, so that even if capitalization slows that growth the overall result is likely to be rising demand for New IT (OECD 1981, Gershuny & Miles 1983, Blackburn *et al.* 1985).

Not only are these sectors growing; precisely because they are, there is a powerful pressure to shift the nature of their delivery, away from a labour-intensive face-to-face basis, towards a technology-based self-service basis (Aglietta 1979, Gershuny and Miles 1983, Blackburn *et al.* 1985). Automatic teller machines, the first experiments in home banking and tele-shopping, the explosive worldwide success of home video rentals, and remote learning systems like the Open University are some of the first expressions. These shifts will result in the growth of production, and of employment, in the advanced information technologies.

Such is the deductive logic. The question is whether the facts support it. In order to reach a verdict, we require data on output and employment growth not just for a few selected advanced industrial countries but for the widest possible range of producers of the New IT. Thus we can hope to build up a global picture of the growth of the new technologies.

The geography of information technology

But there is an additional reason for a broad-based empirical study: to understand the changing international geography of output and employment. There is some evidence, from previous long waves, that downswings were marked by significant shifts in production of leading-edge technologies: witness the growth of the American radio industry in the 1920s and 1930s (Ch. 6), and the development there of wartime radar and computing (Ch. 10), both of which helped lay the foundations for the American primacy in electronics in the post-1950 upswing. Further, there is evidence that new firms, established at this time, can rapidly grow and diversify into complex, multinational organizations, using sophisticated marketing techniques to build and maintain international market power: witness Marconi and RCA during the closing decades of the third Kondratieff, or Apple Computer in the downswing of the fourth.

It is not surprising, therefore, that several recent long-waves studies suggest a model of 'Kondratieff jumping' or 'Technological leapfrogging' between nations, whereby technological leadership passes from one to another (Freeman *et al.* 1982, Ernst 1985, Hoffman 1985, Perez 1985, Soete 1985a). An important conclusion is that several NICs are now well placed, particularly because they have been able to import leading-edge technologies more easily than in the age of American technological monopoly in the 1950s and 1960s; it has been argued that the international technology market has evolved into a more competitive one, just as the Schumpeterian model of technological competitiveness would suggest (Soete 1985a). A contrary view is that this advantage is itself temporary: with increasing automation of IT production, the emphasis will shift back to the advanced industrial countries (Ernst 1985).

Measuring employment change, 1970-81

To examine these questions, we set out to assemble an international data base on the IT supply sector for as many countries as possible. It is not nearly as complete as we would have hoped. It covers only the years 1970–81. It is based on the United Nations International Standard Industrial Classification (ISIC), the only system for which comparative international data can be collected over any extended period. It is not really fine grained enough to pinpoint accurately

Table 11.1 Definition of New IT industries on the basis of the ISIC.

(1) ISIC Group 3825: manufacture of office, computing and accounting machinery	The manufacture, renovation and repair of office machines and equipment such as calculating accounting and punched-card system machines; digital and analogue computers and associated electronic data-processing equipment and accessories; cash registers, typewriters; duplicating machines, except photocopiers; non-laboratory weighing machines; other office machines
(2) ISIC Group 3832: manufacture of radio, television and communications equipment and apparatus	The manufacture of radio, television receiving sets; sound reproducing and recording equipment including public address systems, gramophones, dictating machines and tape recorders; records and tapes; wire and wireless telephone and telegraph equipment; radio and TV transmitting, signalling and detection equipment and apparatus; radar equipment and installations; semiconductor and related devices; electronic capacitors and condensers; radiographic, fluoroscopic and other X-ray apparatus and tubes; parts and supplies specially used for electronic apparatus classified in this group
(3) ISIC Division 385: manufacture of professional, scientific, controlling and measuring equipment (NES); photographic and optical and time-keeping goods	Group 3851 covers the manufacture and repair of laboratory and scientific instruments and of measuring and controlling equipment not elsewhere classified; the production of accelerators; medical and dental instruments and equipment. Group 3852 covers the following: the manufacture of optical instruments and lenses, ophthalmic goods; photographic and photocopying equipment and supplies. Group 3853 covers the manufacture of watches and clocks of all kinds and other timing devices

the New IT industries: the best we can do is to use a combination of its three-digit and four-digit categories (Table 11.1).

This definition thus consists of three major subcategories. The first, Group 3825, covers all kinds of electronic and electro-mechanical office machinery and equipment from typewriters to computers. The second, Group 3832, contains a very wide range of New IT activities including semiconductors and other components, telecommunications equipment, electronic consumer goods such as radios, TV and sound reproducing systems, and other electronic capital goods such as radar and broadcasting apparatus. It thus contains many key elements of the commodity groups that are defined today as New IT and in employment terms it is more important than the other two subcategories. The third, ISIC Division 385, is a more miscellaneous category that embraces professional and scientific measuring and control instruments and equipment along with photographic and other optical goods as well as clocks, watches and other timing devices.

For these categories we assembled a 38-country employment data base for 1970–81. It was based on UN data, supplemented by additional data obtained from national statistical offices and industry associations, OECD and EEC statistics and trade information. It covers all OECD countries, a large number of NICs (Argentina, Brazil, Hong Kong, India, Indonesia, Malaysia, Mexico, Philippines, Puerto Rico, Singapore, South Korea and Taiwan) and two Comecon countries, Poland and Yugoslavia.

From 1970 to 1981, aggregate New IT employment in the 38 countries increased significantly, from 5 349 000 to 6 863 000, or 28%. This appears to confirm the hypothesis that such leading-edge industries will run counter to the prevailing trend during a Kondratieff downswing. For 32 of these 38 countries, we could calculate total manufacturing employment: it grew from 82.0

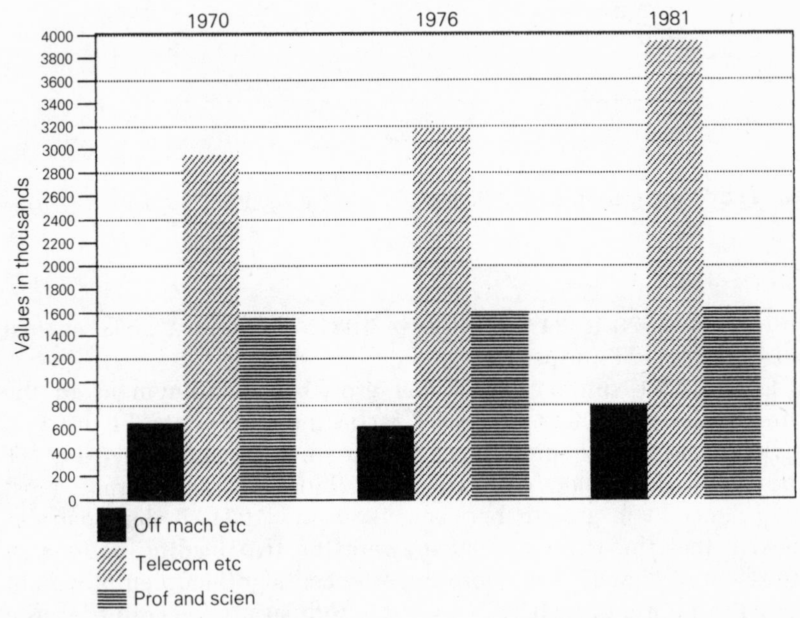

Fig. 11.1 Employment change, New IT industries, world estimate, 1970–81
Note: 'World estimate' refers to 39 leading countries for which data were available.

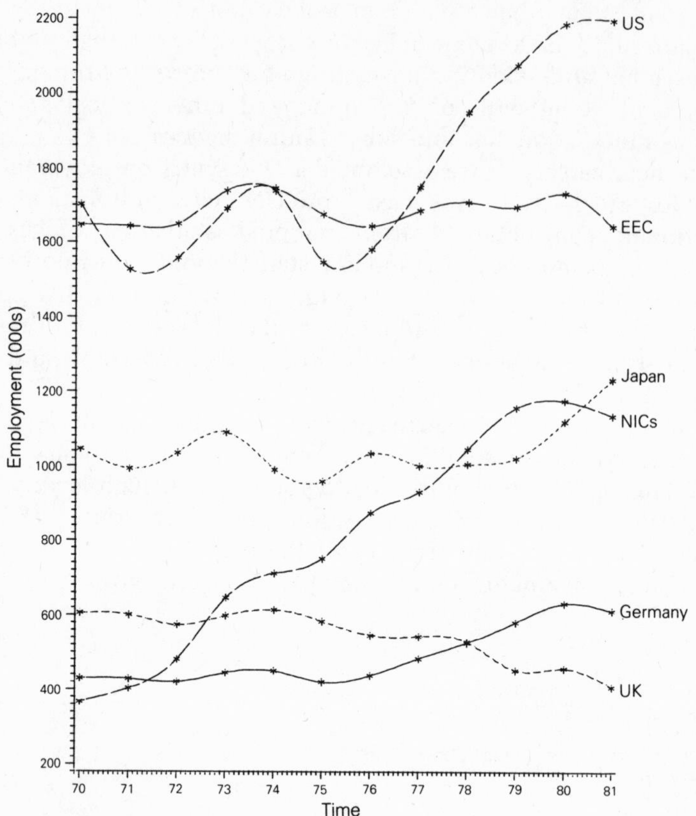

Fig. 11.2 Employment, New IT industries, selected countries, 1970–81

million in 1970 to 84.9 million in 1981, or by only 3.5%, against 28% for New IT industries.

Figure 11.1 summarizes global growth of employment in the three major subsectors of New IT according to our data. Figure 11.2 gives a summary picture of employment trends in New IT industries in major countries and blocs, showing that most experienced job growth between 1970 and 1981. It demonstrates clearly that the USA and Japan were the two leading nations in numbers of New IT jobs; both experienced significant employment growth in these industries. The newly industrializing countries as a group also experienced a very steep rise in employment, albeit from relatively low initial levels (though lack of statistics in some may

Table 11.2 New IT: top 12 nations ranked by absolute employment growth, 1970–81.

	1970 employment	1981 employment	Absolute job growth 1970–81
USA	1 702 000	2 198 000	496 000
Taiwan	60 000	293 870	233 870
Japan	1 045 000	1 233 000	188 000
West Germany	429 000[a]	612 000	183 000
South Korea	24 500	183 200	158 700
Hong Kong	45 700	140 700	95 000
Singapore	12 140	71 420	59 280
Malaysia	17 000	75 700	58 700
Poland	139 000	185 000	46 000
India	77 000	122 000	45 000
Brazil	106 800	150 600	43 800
Philippines	5200[b]	39 600	34 400

Source: RUIT data base.
Notes:
[a] Particular difficulties were experienced in obtaining data for West Germany and fitting them into the ISIC classification.
[b] 1972; no data for Philippines prior to this date.

tend to underestimate early levels). In striking contrast, the EEC bloc experienced no employment growth in New IT, though some member countries, especially the Federal Republic of Germany, did so. The UK suffered a dramatic decline in New IT employment, from about 600 000 to 409 000, or almost one-third; more finely disaggregated national data show a decline only from 561 000 in 1971 to 492 000 in 1981, about 12.3%. Clearly, whatever the precise values, both sets of data indicate a relatively poor performance.

In summary, then, as measured by direct job-creation in New IT industries the USA, dominant at the start, kept a commanding lead; Japan kept a strong second place; the EEC countries performed relatively poorly, and in the UK's case very poorly; the NICs grew impressively, though in 1981 their combined New IT employment remained below that of either the entire EEC bloc or Japan.

Table 11.2 shows the detail of this picture, by ranking the 12 leading nations in terms of absolute job gains over the 1970–81 period. The USA and Japan appear in first and third places, with 496 000 and 188 000 extra jobs respectively; an NIC, Taiwan, edges into second place with 234 000. Altogether, of the top 6, 3 are AICs,

Table 11.3 New IT: top 12 nations ranked by percentage employment growth, 1970–81.

	1970–81		1976–81	
	% per year	Absolute (10–year)	% per year	Rank
Indonesia	30.0	16 150	21.3	2
South Korea	20.1	158 700	1.9	17
Singapore	17.5	59 280	8.7	8
Taiwan	15.5	233 870	10.3	5
Malaysia	14.5	58 700	17.6	3
Hong Kong	10.8	95 000	8.9	7
Israel	9.7	16 540	9.1	6
Turkey	9.3	6300	1.5	20
New Zealand	7.6	3370	−0.5	26
Puerto Rico	7.4	9390	6.3	12
Mexico	6.0	10 520	8.0	9
Finland	5.9	7400	0.1	24

Source: RUIT data base.

Notes:

UK New IT industry reported a mean annual job loss of 3.5% over this period; the USA experienced a 2.4% mean annual rate of job growth and Japan 1.5%. All but 6 of the 39 countries covered experienced some employment growth in the new IT industries.

For 1976–81 the ranking of the 12 countries with the highest mean annual exployment growth rates was: Philippines 28.3% (1972–81), Indonesia 21.3%, Malaysia 17.6%, Ireland 13.7%, Taiwan 10.0%, Israel 9.1%, Hong Kong 9.0%, Singapore 8.8%, Mexico 8.0%, West Germany 6.9%, USA 6.8%, Puerto Rico 6.3%.

3 NICs; of the top 12, 8 are NICs and one is a Comecon country. This impressive performance by the NICs emerges even more sharply from Table 11.3, which ranks countries by percentage growth over the period 1970–81. The Asian NICs emerge outstandingly at the top, though – since many started from such small numbers – their aggregate gains were less impressive; three however, Taiwan, the Republic of Korea and Hong Kong, even posted large absolute gains. The data also shows that after 1976 individual Asian NICs posted sharply divergent performances and that two advanced countries, the USA and West Germany, appeared among the twelve fastest-growing nations.

Another measure of performance is a relative one: how strongly a nation has concentrated its economic structure on the leading-edge industries. Geographers commonly employ the Location Quotient (LQ) as a measure of such concentration. In Table 11.4 the LQ is used to measure the relative importance of New IT employment to

Table 11.4 New IT: top 12 nations ranked by location quotient, 1970, 1976, 1981.

	1970	1976	1981
Singapore	1.49	3.25	3.11
Hong Kong	1.28	1.72	1.92
Malaysia	0.80	1.43	1.87
USA	1.43	1.30	1.45
Japan	1.37	1.34	1.44
Puerto Rico	0.88	1.27	1.40
Switzerland	1.89	1.51	1.18
Israel	0.68	0.96	1.15
South Korea	0.44	1.14	1.11
West Germany	0.80	0.89	1.07
Ireland	0.85	0.76	1.01
UK	1.16	1.08	0.87

Source: RUIT data base (UN, various national industrial statistics etc.).

Note: The location quotients are calculated from data for employment in New IT industries and in all manufacturing for as many countries as possible; because of data difficulties, 6 of the 39 countries under study had to be excluded from this exercise; they were Argentina, Brazil, Mexico, Philippines, Taiwan, Thailand.

all manufacturing employment for 33 of the 38 countries in the data set (the remainder lack necessary data); an LQ above 1.00 indicates an above-average concentration of New IT jobs. As in previous tables, the 12 leading nations are ranked. By 1981 three NICs, Singapore, Hong Kong and Malaysia, occupied the top three places, with the USA and Japan in fourth and fifth places respectively. The three NICs had sharply increased their new IT concentration over the preceding decade; the USA and Japan recorded much smaller shifts. The most striking case among the 12, in some ways, was the UK: the only one to record a shift away from New IT employment over the period.

Changes in individual New IT categories

Figures 11.3–11.5 present data for the three main constituents of New IT as we have defined it. For comparisons between countries at any point in time, the data on which they were based present even more problems, if anything, than the aggregate data: some countries lacked data, in all there were problems arising from rapid technical change and consequent classification boundary shifts. But, in terms of indicating relative changes between countries over time, they do tell a reasonably consistent story.

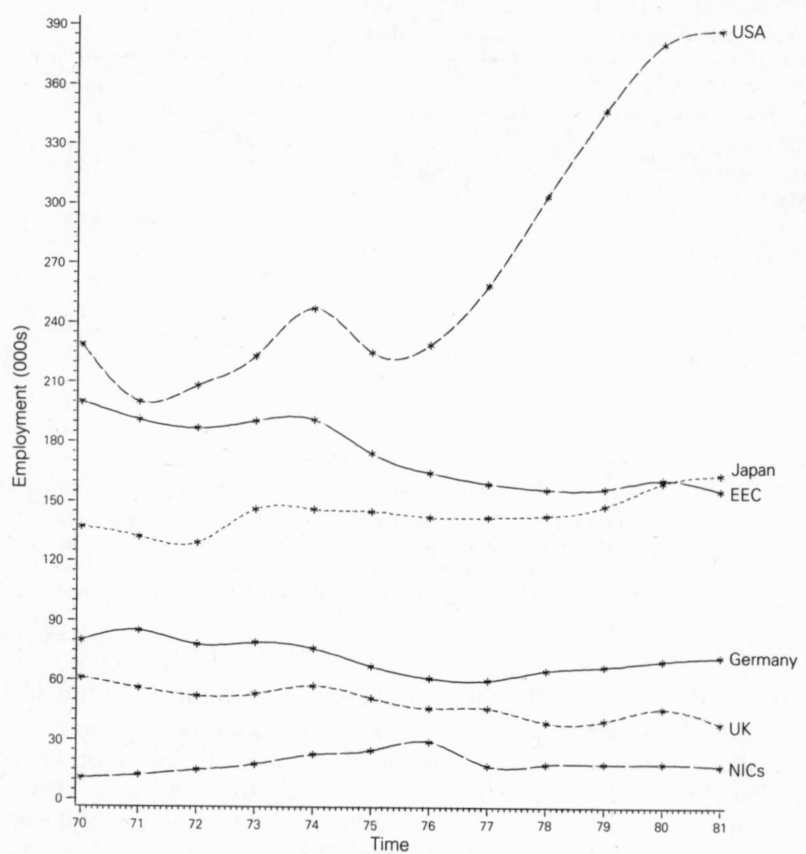

Fig. 11.3 Employment, office machinery, selected countries, 1970–81

Office machinery It proved possible to obtain separate data for only 27 of the 38 nations. Figure 11.3 shows that the USA maintained a clear leading position, with huge gains after 1975. Japan and the EEC bloc came next, with Japan pulling ahead after 1980, but in comparison to the USA neither registered large gains; indeed, the EEC lost jobs over most of the period. The UK had a particularly marked record of employment losses. Neither did the NICs perform impressively, though this may reflect the omission of both Taiwan and Singapore. So this was overwhelmingly a sector dominated by the Americans in this period, which reflects the fact

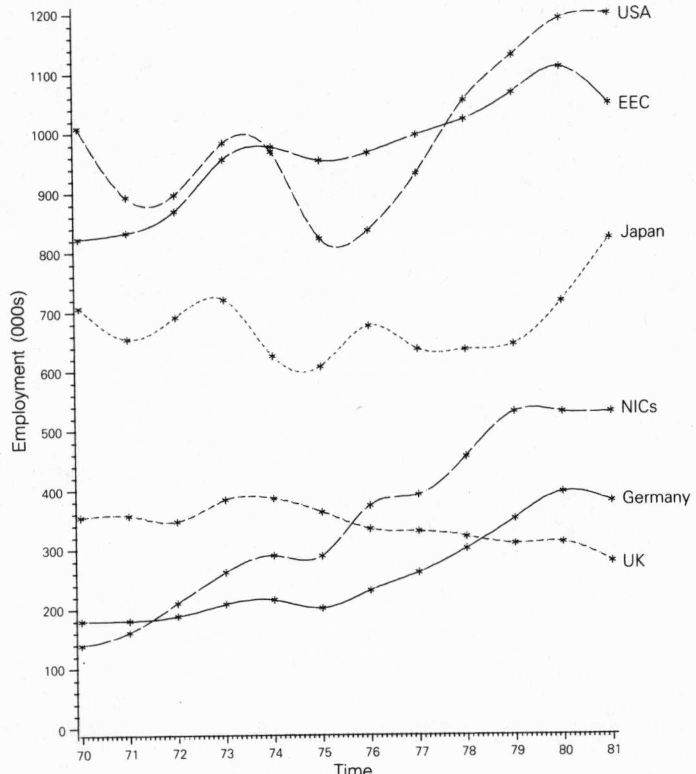

Fig. 11.4 Employment, telecommunications, components, radio and TV equipment, selected countries, 1970–81

that it includes the fast-growing computer sub-sector.

Telecommunications, components and consumer electronics This is the largest sector in terms of total employment. Figure 11.4 shows that in 1981 the USA led the EEC bloc by a small margin after dipping below it in the mid-1970s. The EEC had enjoyed fairly steady growth except for 1980–1; both the USA and Japan demonstrated a more volatile pattern, albeit with strong growth in the late 1970s. Again, there were major differences within the EEC: the Federal Republic demonstrated fairly steady growth, while employment in the UK declined by 80 000, or about 22%, over the period. The NICs demonstrated substantial growth – unsurprisingly, since this category includes the consumer electronics and semi-

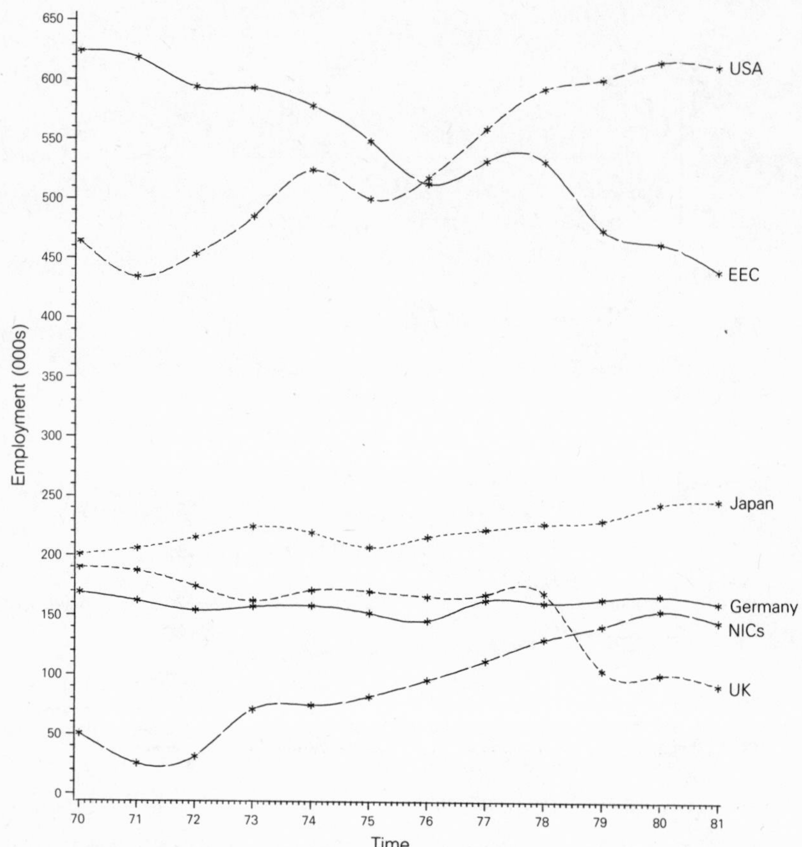

Fig. 11.5 Employment, professional and scientific instruments, etc., selected countries, 1970–81

conductor-based activities that have been located there in such strength. But still in 1981 the 12 countries in this group accounted for only some 17% of the combined employment of the USA, the EEC and Japan: a useful corrective to those who may think that they diverted large numbers of jobs from the advanced industrial world.

Professional and scientific instruments etc. Employment in this category, the second largest in employment terms in 1981, had grown only slowly since 1970, from 1.56 to 1.66 million. Figure 11.5 shows that the USA and the EEC had the lion's share of this employment. But the EEC, which had a clear lead in the early 1970s, showed a sharply declining trend and was overtaken by the rapidly growing American economy after 1976; thus, over the period the rôles were almost precisely reversed. Some of the EEC loss is explained by the UK, which suffered particularly serious losses in 1978–9 for reasons that are not clear. Japan grew modestly, but far more slowly than in the other two sectors; the NICs grew more strongly after a dip in the early 1970s, almost tripling employment over the period.

Though the trends differ in detail, there are strong common features: the strong growth of the USA and Japan, the relatively weak performance of the EEC save in the telecommunications sector, the marked contraction of the UK in all three areas, and the steady expansion, office machinery excepted, within the NICs. The conclusion seems inescapable: though New IT did indeed generate large numbers of additional jobs worldwide during the fourth Kondratieff recession, the geography of that employment was showing a major geographical shift, out of Western Europe, into Japan and the Asian NICs. Only the American dominance provided a relatively fixed point: around it, the centre of gravity was shifting from the Atlantic to the Pacific.

Employment and output in the USA, the UK and Japan, 1958–82

Though this evidence is striking, it covers all too short a period of time. Nor does it relate to changes in output. For three of the leading industrial economies – the USA, the UK and Japan – we can, however, do better, taking the analysis back to 1958 (1963 for American output): a period that includes much of the Kondratieff upswing.

Figure 11.6 shows the results for employment. All three countries recorded consistent job growth from the late 1950s to the late 1960s, but thence the patterns diverged: British employment declined, at first gently, then (after 1980) steeply; the USA and Japan showed considerable gains, albeit with much fluctuation. This underlines the point that the British experience is highly anomalous: it cannot be satisfactorily explained by worldwide trends, whether in terms of a Kondratieff downswing, or of a new international division of labour between the AICs and the NICs. It seems to represent a unique competitive failure.

The point is further emphasized by an analysis of output growth (Fig. 11.7). The problems of comparison are even more severe than

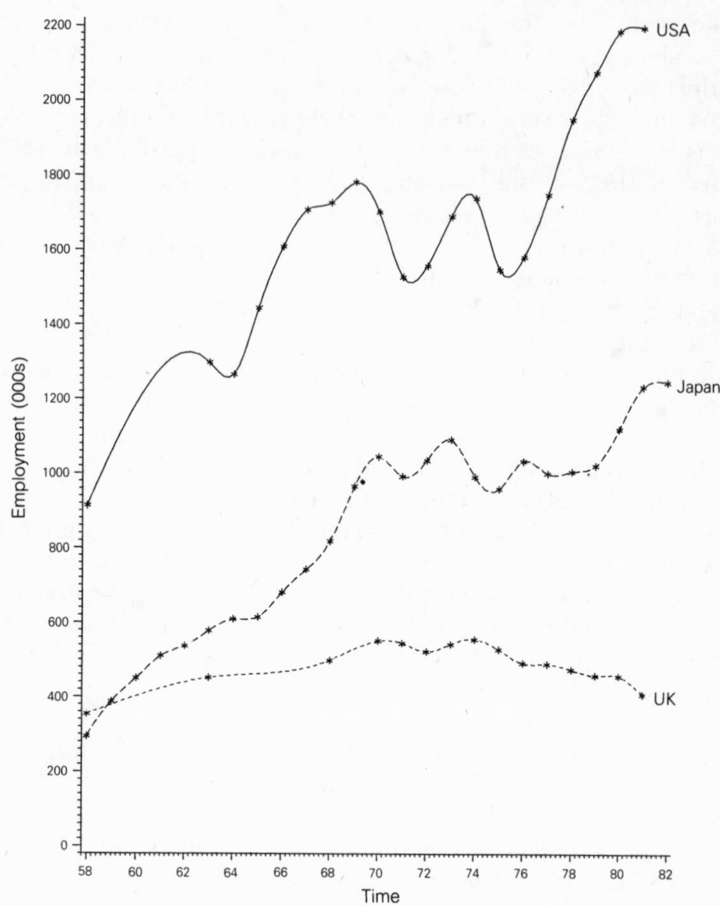

Fig. 11.6 Employment, New IT industries, USA, UK, Japan, 1958–82

in the case of employment. Nevertheless, the estimates suggest a clear and widening divergence of output growth between the USA and Japan on the one hand, the UK on the other. This is particularly marked after 1974, when British output stagnated while output in the other two economies began to grow sharply.

Overall, this extended analysis deepens our understanding of the processes of international competition over the fourth Kondratieff. In its upswing phase the British economy already displayed

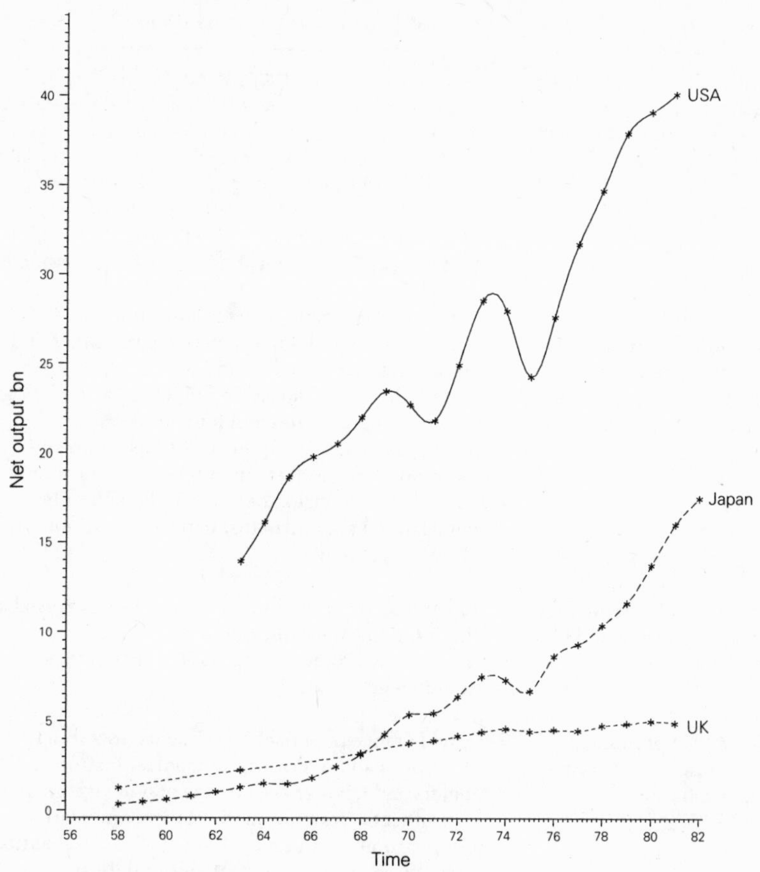

Fig. 11.7 Net output, New IT industries, USA, UK, Japan, 1958–82
Note: Output valued in pounds sterling (1980 prices).

abnormally weak growth of both employment and output in the New IT industries in relation to the economy that then offered the most apt comparison, that of the USA. At this point Japanese employment and output were already rising rapidly, about as fast as the American. In the downswing, the UK showed stagnant output accompanied by declining employment, consistent with productivity gains through process innovation; in the other two economies, too, output rose much faster than employment, but the critical point is that both were still growing strongly. The

Table 11.5 New IT trade product groups: definition.

Division 75: office machines and automatic data-processing equipment	Typewriters; calculating and cheque writing machines; computers and other automatic data-processing machines and peripherals for same; cash registers incorporating a calculating device, postage, franking and ticket issuing machines incorporating calculating devices; duplicating machines; photocopying and thermo-copying machines; parts for this type of equipment not elsewhere classified
Division 76: telecommunications and sound recording and reproducing apparatus and equipment	Television and radio broadcast receiving equipment; gramophones, dictating machines and all other sound recorders and reproducers, including record players and tape decks and systems; microphones and loudspeakers etc; telephonic and telegraphic apparatus; radio and television transmitters; other telecommunications equipment; television cameras; radar and radio navigational aid apparatus; parts for these not elsewhere classified
Group 776: electronic tubes, valves, transistors and other semiconductors	Electronic valves and tubes; piezo-electric crystals; diodes; transistors and similar semiconductor devices; electronic microcircuits; parts of these components
Division 87: professional, scientific and controlling instruments etc.	A wide range of optical instruments, including telescopes and microscopes; medical instruments; meters and counters; surveying and navigational instruments; drawing, mathematical and marking out instruments and machines; electronic, electrical and other automatic control and analytical instruments; parts for these products not elsewhere classified
Division 88: photographic apparatus and supplies; optical goods; watches and clocks	Photographic and cinematographic cameras, projectors, sound recorders, etc; photographic film, plates and paper; optical goods including lenses, spectacles, etc; watches, clocks and other time-recording equipment, time switches, etc.

conclusion must be that – just as most long-wave theorists have surmised – Kondratieff downswings are indeed everywhere characterized by process innovation and strong improvements in output : employment ratios; But what is striking and anomalous is the weakness of the British industrial performance in these leading-edge technologies, whether measured by employment or by output.

International trade in New IT

Another way of measuring relative international performance and competitiveness is through trade, especially export performance. Accordingly, we have tried to measure the leading industrial nations' share of 'world' exports of New IT products. Data here are at first sight more complete and detailed than for output or employment. But they still present problems: discrepancies between countries in the way imports and exports are counted, currency fluctuations, and – a special problem for our industries – exportation for processing followed by reimportation. These need to be borne in mind in what follows.

Categories included in our analysis are shown in Table 11.5; they are based on the standard international trade classification employed by the UN, OECD and other international agencies; there are four two-digit and one three-digit headings, which correspond quite closely to the ISIC product categories just used in our study of employment and output changes. Data were gathered for the same group of 38 countries earlier analysed, with the exception of Puerto Rico and (for Category 776) Taiwan and Poland.

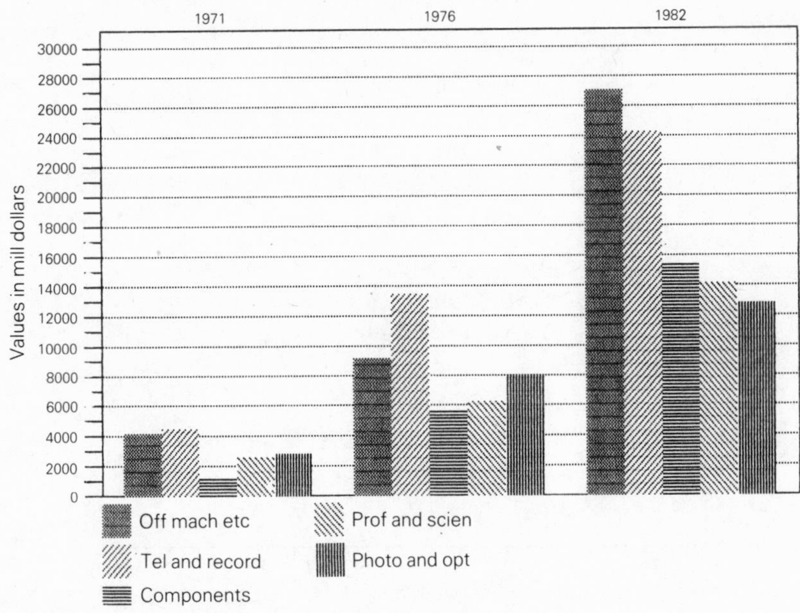

Fig. 11.8 Exports, New IT, world estimate, 1971–82
Note: 'World estimate' refers to imports to all OECD countries.

Over the period 1971–82, the value of New IT exports rose considerably worldwide, from $15,985 million to $94,334 million at constant 1980 prices. Figure 11.8 shows that all the separate categories shared in this growth. Against this background, Figure 11.9 demonstrates that in 1971 the EEC bloc had a clear lead, commanding approximately 45% of the exports of the entire group of trading countries. But this lead was steadily eroded during the 1970s as the USA, Japan and the NICs increased their export shares; by 1982 the EEC's total share was only 32%, against the USA's 24.5% and Japan's 21.8%. Within the EEC bloc the Federal Republic maintained its position as third trading nation, though its share fell from 15.9% to 10.8%. The UK, in fourth place, similarly saw its share fall from 7.9% to 5.6%; France, Italy and the Netherlands similarly recorded a drop in export share, as indeed did all the European countries save Ireland, Spain, Austria and Finland which

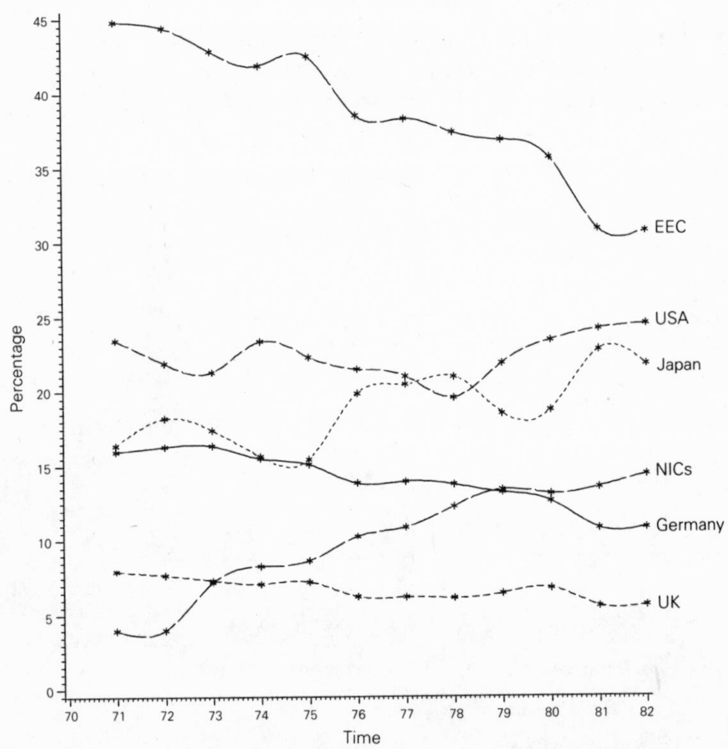

Fig. 11.9 Exports, New IT, per cent of estimated world total, 1971–82
Note: 'Estimated world total' refers to imports to all OECD countries.

had been very minor producers in 1971.

The most spectacular export growth came from the NICs in the group, which increased their export share from 4% to 14%; among them, Hong Kong, Singapore, the Republic of Korea and Taiwan all had notable growth, and Hong Kong's 2.7% share in 1982 was almost half the British figure. Yet still, in 1982 their combined export share was a long way behind the market leaders. The USA and Japan had 38% of total exports in 1971; in 1982 they had 46%. Clearly, the leading advanced nations were not yet ceding their position to the NICs; rather, the weaker members of the group – above all the Europeans – were the losers.

Analysis of the individual categories brings out some significant variations. In *computers and other office machinery* (Fig. 11.10), the largest export category in the early 1980s, the EEC share fell from 47% to 32% between 1971 and 1982; the USA increased its share from 33% to 36%, thus coming to dominate export markets; Japan doubled its share from less than 8% to over 15% (Table 11.6), with major export achievements in such products as typewriters, copiers and computer hardware; the NICs doubled their share, which still however remained modest by 1982. These figures do not tell the whole story, since the market leaders also produce in their export markets: thus as much as three-quarters of British computer exports in the early 1980s were said to be from American subsidiaries (National Research Council 1984), and in mid-1985 the Japanese were just starting typewriter production in the UK to avoid antidumping duties. Overall, a striking feature is again the

Table 11.6 Office machinery etc. (SITC 75): leading export nations (ranked by 1982 shares of 'world' exports).

	1971 share (%)	1976 share (%)	1982 share (%)
USA	33.57	32.47	36.60
Japan	7.82	10.65	15.53
Germany	15.76	13.42	10.47
UK	8.13	8.81	7.32
France	10.23	9.03	5.75
Italy	7.86	5.94	4.25
Netherlands	3.70	3.19	3.34
Canada	3.23	3.16	2.69
Ireland	0.16	1.09	2.48
Sweden	3.47	2.92	1.79

Source: OECD trade statistics.

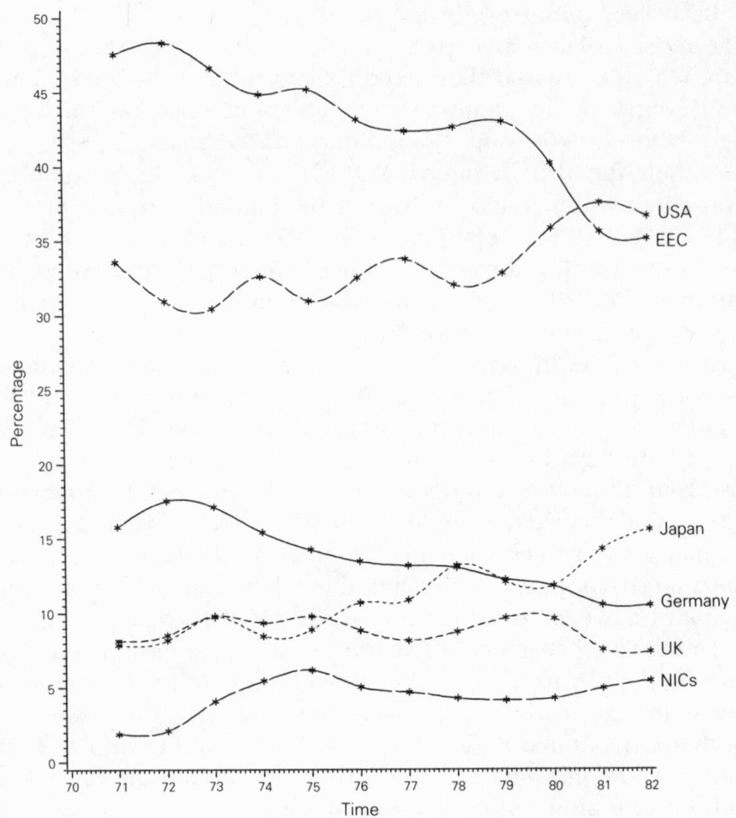

Fig. 11.10 Exports, computers and other office machinery, per cent of estimated world total, 1971–82

Note: 'Estimated world total' refers to imports to all OECD countries.

relative decline of the major European producers, including not merely the UK but also Germany, France, Italy and Sweden (Table 11.6).

In *telecommunications and sound recording etc. equipment* (Fig. 11.11), Japan increasingly dominated export trade over the whole 1971–82 period: during the 1970s it held between 28% and 36% of total trade, but by 1981 it had 40%, thence falling marginally to 38% in 1982. In comparison the USA had a marginally declining 8–9% share over the period. The EEC held its share until the mid-

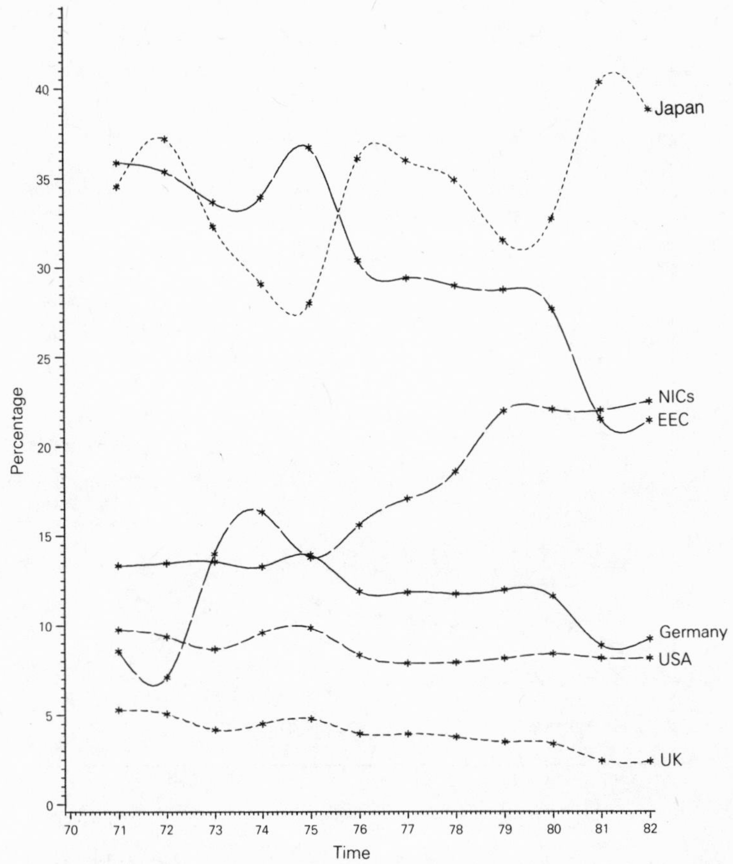

Fig. 11.11 Exports, telecommunications and sound recording apparatus, per cent of estimated world total, 1971–82

Note: 'Estimated world total' refers to imports to all OECD countries.

1970s, but then declined precipitately; this equally affected stronger performers like the Federal Republic and weaker ones like the UK. Conversely the NIC group recorded a strong advance in the late 1970s, rising from 14% to 23%; five of them (Taiwan, the Republic of Korea, Mexico, Singapore and Hong Kong) appeared among the top ten exporters of 1982.

This category is particularly interesting because it includes many

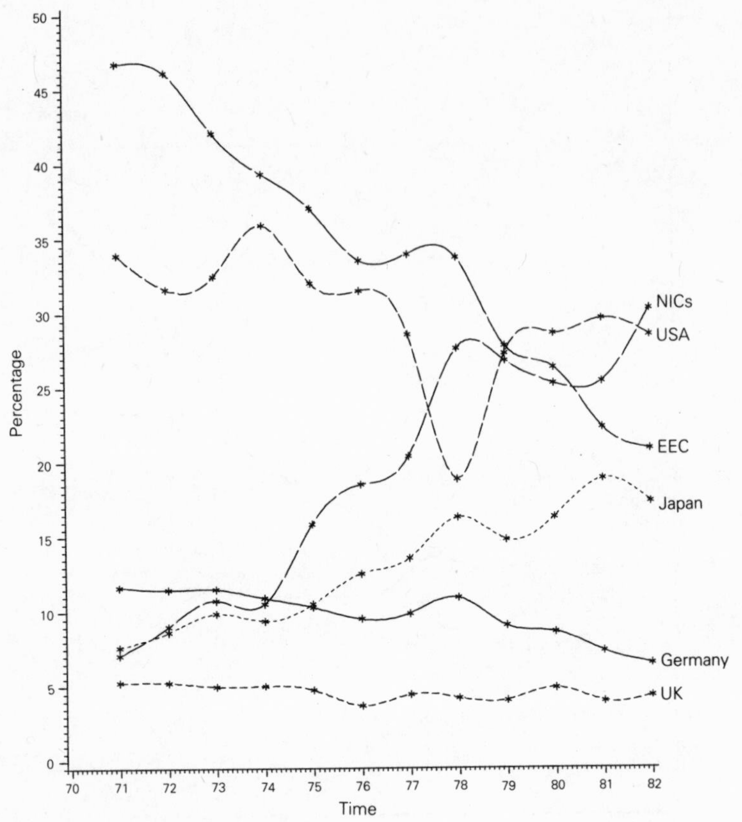

Fig. 11.12 Exports, electronic components, semiconductors, etc., per cent of estimated world total, 1971–82

Note: 'Estimated world total' refers to imports to all OECD countries.

relatively mature products which might be expected to be more open to diffusion. In the mid-1950s the USA had twenty times Japan's exports in this group; by the early 1980s Japan had twice the USA's (National Research Council 1984). Japan began to play a major rôle in consumer radios and audio tape recorders in the 1950s. In colour television, where the Americans had played the crucial innovatory rôle, Japan began to displace it in the early 1970s; using the latest components, exploiting product reliability

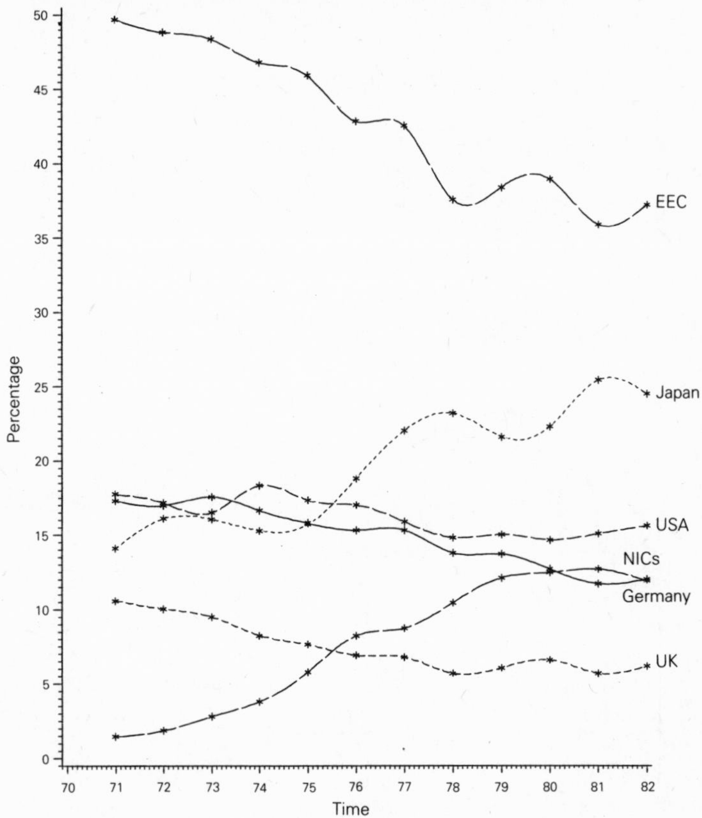

Fig. 11.13 Exports, photographic, etc., apparatus, per cent of estimated world total, 1971–82

Note: 'Estimated world total' refers to imports to all OECD countries.

and automated production techniques, it captured over half world production and three-quarters of exports by 1977, a share that thence declined only because its producers established production facilities in major export markets.

The third category, *electronic components, semiconductors, etc.*, was the third largest in exports in 1982 (Fig. 11.12). Particularly notable are the sharp increase in the NICs' market share, a smaller but still perceptible increase in that of the Japanese, a relative fall

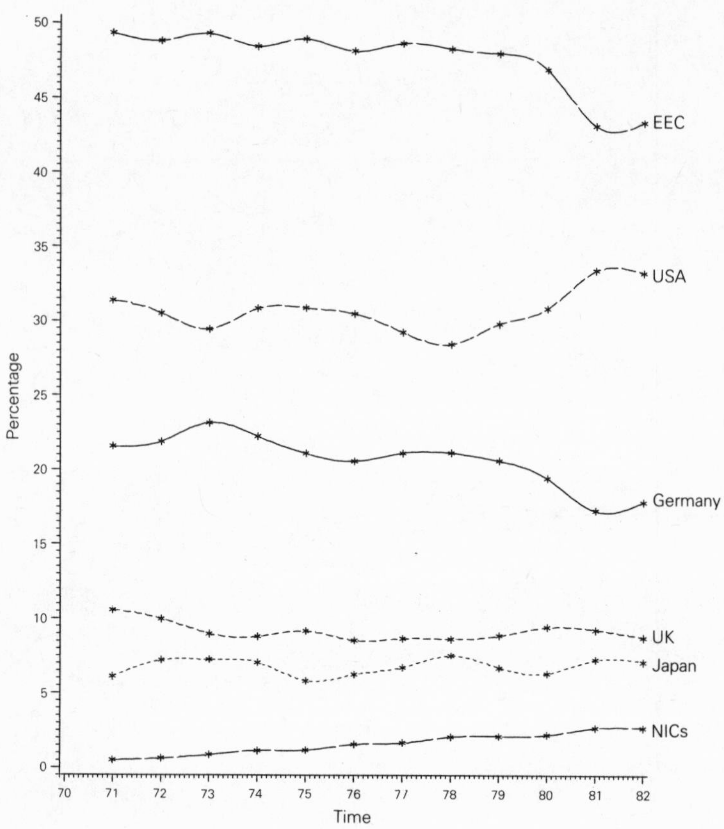

Fig. 11.14 Exports, professional and scientific instruments, etc., per cent of estimated world total, 1971–82

Note: 'Estimated world total' refers to imports to all OECD countries.

in that of the USA and a steeper one in that of the EEC. This, curiously, is the only category in which the UK even approximately retains its market share.

In *photographic apparatus, optical goods, watches, etc.*, a category embodying increasing amounts of New IT, the USA maintained a steady share, the EEC sharply lost, and both Japan and the NICs gained (Fig. 11.13). In *professional and scientific instruments*, the picture is different: though here too the EEC bloc lost share and the USA marginally gained, the other protagonists (Japan and the NICs) failed to achieve substantial increases in market share (Fig. 11.14).

Table 11.7 New IT exports: percentage composition by sector, 1982.

	UK	USA	Japan	FRG	EEC	NICs	Total (38 nations)
office machinery, ADP equipment	38	43	20	28	33	11	29
telecommunications and sound recording	11	9	46	22	16	41	26
transistors, valves, etc.	13	19	13	10	11	35	16
professional and scientific	24	21	5	25	21	3	15
optical goods etc.	14	8	16	15	19	10	14
total	100	100	100	100	100	100	100

Source: UN and OECD trade statistics.

Table 11.8 New IT: net trade balances, 1982.

Rank		Net trade balance, 1982 ($m, current prices)
1	Japan	17 345
2	USA	4794
3	Taiwan	1323
4	Malaysia	620
5	South Korea	256
6	Ireland	139
7	Switzerland	42
8	Philippines	−22
9	Poland	−29
10	Mexico	−149
38	UK	−3799

Source: UN and OECD trade statistics.

There is another way of analysing these figures: by asking what proportion of a country's total New IT exports is represented by each category. Table 11.7 summarizes the results. Office machinery, including computers, dominates American IT exports, with 43% of the total, and, perhaps more surprisingly, that of the UK, with 38%; it made up one-third of the EEC total, only one-fifth of Japan's, and a mere 11% of the NICs'. Consumer electronics, including telephone equipment, dominated the export performance of Japan, with almost half the New IT total, and of the NICs, with 41%; in sharp contrast it accounted for only 11% of British exports and 9% of American – a reflection, in part, of the American tradition of overseas manufacture of telephone equipment through subsidiaries. Components dominated the export performance of the NICs, with 35%. The picture conforms to a model of an international division of labour, in which the USA and UK concentrate on the more sophisticated computer products, while Japan concentrates on the consumer sector and the NICs supply components.

All industrial countries import New IT products as well as exporting them; net trade balances summarize their overall trading performance (Table 11.8). Japan emerges as outstandingly most successful, with a net trade surplus far ahead of the USA, its nearest competitor. Particularly notable is the weak performance of the UK, which appears at the very bottom of the list with a huge net deficit in New IT products – further confirmation of its structural problem in moving into leading-edge technology, which has emerged repeatedly in the course of this analysis.

Summing up

The picture that has emerged from this analysis, we would argue, is a very consistent one. Over the fourth Kondratieff the USA early assumed a commanding lead in New IT production, which it successfully maintained until the early 1980s. The degree of its success in international markets cannot adequately be measured by trade statistics, because in key areas (e.g. telecommunications, computing equipment) it exported its products by manufacturing them through subsidiaries in its competitor countries. It proved particularly strong in the most advanced and sophisticated technologies, such as computing. It achieved large increases in both employment and output not only during the Kondratieff upswing but also – after a distinct setback in the early 1970s, and not without perturbations thereafter – during the subsequent Kondratieff recession. Though American output of New IT climbed faster than employment, indicating productivity gains through the application of technology to production, the gain in jobs, even during the recession, was impressive.

Japan's performance is equally impressive, not least for its relative steadiness. It started from a far weaker base in New IT than the USA, but steadily narrowed the gap. It too continued to expand output and employment during the recession, again with some fluctuations, and with increasing productivity gains. It proved outstandingly successful in the export-oriented consumer electronics sector, where it eventually beat the USA into top place.

In contrast the performance of the world's third great industrial bloc, the EEC, reveals serious competitive weaknesses. It joined in the general expansion of output and employment during the upswing years of the late 1950s and 1960s, but then went into relative decline. It lost its position in the important office products market to the USA. Its performance in the other two major New IT sectors was equally lacklustre. Within the EEC, the Federal Republic's performance was clearly better than average; the UK's was on every measure much worse. Every piece of evidence, analysed in this chapter, confirms that the UK was failing to compete in the production and export of the New IT products.

The extent of this failure has been assessed, and the reasons for it convincingly analysed, in a recent study by Ian Mackintosh. He concludes that the quarter–century from 1960 saw 'the virtual collapse of Europe as a tenable competitor in the most important industrial technology of the century' (Mackintosh 1986, 70). He ascribes this failure to a number of causes: an extraordinary delay in appreciating the revolutionary nature of silicon technology and the integrated circuit, perhaps because of geographical remoteness from the American centres of innovation; lethargy on the part of management in grasping the significance of new cost–reduction techniques; the political failure to create a common market in high–technology products; the preference of European scientists for fundamental over applied research; the reluctance of European financiers to supply venture capital; and, finally, the inability to create a geographical 'skill cluster' of the order of a Silicon Valley (Mackintosh 1986). The result, he makes plain – thus independently confirming our own conclusions – is that Europe 'has failed abjectly to keep up with its main international competitors, the US and Japan, and it is now further threatened by new and determined competitors located in South–East Asia' (ibid., 83).

The counterpart of the failure of the EEC, and above all of the UK, was in fact the success of the NICs, notably those of Eastern Asia. Essentially starting late, at the very end of the Kondratieff upswing, they sharply expanded both production and employment throughout the recession. Their trade performance was especially impressive in the less sophisticated products for which they could acquire the necessary technological capacity from the more advanced countries; by the early 1980s they had still not made major inroads in the more sophisticated office equipment and computing sector.

All this leads to two emphatic conclusions. Firstly, it does prove true that, on the world scale, leading-edge industries will continue to expand through a Kondratieff recession, thus laying the foundations for the subsequent upswing. Secondly, such expansion may be marked by maintenance of market share on the part of the most powerful established producer nations, coupled with quite profound and rapid shifts in fortune among others. In particular, newly industrializing nations may make surprisingly rapid inroads in relatively advanced technologies, whereas old-established and technologically sophisticated countries may suffer drastic failures of competitiveness. That last conclusion applies with especial force to the UK, whose problem we now go on to analyse in detail.

12
The British anomaly

We focus now on the paradox of the relative decline of British New IT during the fourth Kondratieff. Chapter 11 has suggested that starting from an apparently strong base after World War II, the sectors constituting New IT in the UK suffered from abnormally slow growth of employment and output, giving way in the 1970s to actual contraction of jobs, and accompanied by a rapid deterioration in the country's trading position in these important sectors. Far from representing part of the solution to the country's economic problem, they seem to have become part of the problem.

This appears to be a peculiarly British problem. As already discussed in Chapter 11, and earlier in this book, students of Kondratieff long waves have vigorously debated their significance for job creation and job destruction. One view, initially expressed by Freeman *et al.* (1982), is that large-scale employment growth is characteristic only of the first product-innovation phase of the Kondratieff; in a second, output grows but employment stagnates ('jobless growth') as process innovations begin to predominate; finally, output slows and employment declines. Freeman's group have identified these phases for manufacturing in the EEC countries during the fourth Kondratieff; the first, they suggest, ran from 1950 to 1965; the second from 1965 to 1973–4; the third from 1974 (Freeman *et al.* 1982, Rothwell 1982).

This is why the experience of British New IT has particular significance. For Freeman has suggested that even a leading-edge industry like electronics may follow the same pattern (Freeman *et al.* 1982). But this conclusion may follow from the behaviour of the EEC economies and particularly of the UK, which prove to be anomalous in a worldwide context. There, our analysis in Chapter 11 has suggested that leading-edge industries do indeed move counter to the general trend, generating additional employment as well as additional output throughout the Kondratieff recession phase. The problem then becomes to account for the divergent British experience.

In this chapter, therefore, we first describe that experience in greater detail. Then we go on to discuss possible explanations.

Employment and output in the third and fourth Kondratieffs

Even in a country with a long record of detailed data collection such as the UK, attempts to create long time series for a particular new technology sector may be thought 'heroic by any standards' (Soete & Dosi 1983, 5; cf. NEDO 1967). But we believe that by combining and cross-checking the censuses of population and of production it is possible to do so. There are many problems: Census of Production data for industries begin only in 1907, and before that refer to occupations; the data tend to fail to identify New IT industries until they are well established. But we think that we have overcome these problems sufficiently to give a long-term overview since the start of the 20th century.

Figure 12.1 summarizes the results. It shows fairly consistent growth in the New IT industries in the first two decades of the 20th century – indeed, possibly underestimated because of the identification problem just mentioned – which accelerated from about 1930,

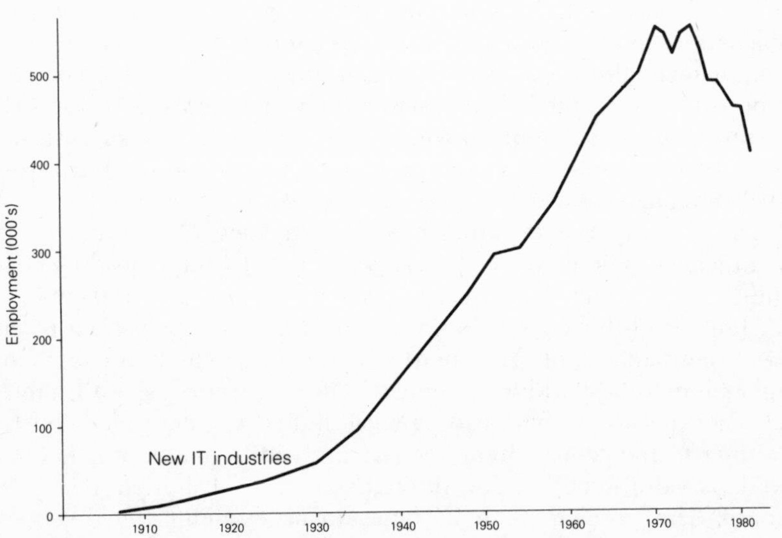

Fig. 12.1 Employment, New IT, UK, 1907–81

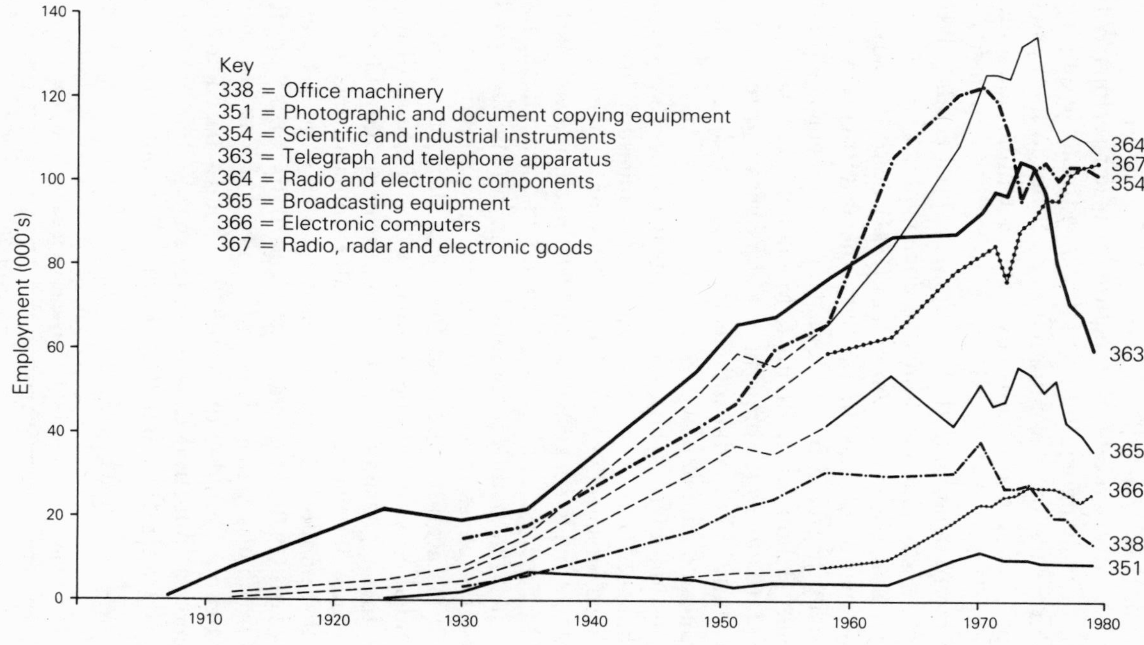

Fig. 12.2 Employment, constituent New IT industries, UK, 1907–79

the time of the third Kondratieff depression. Throughout the fourth Kondratieff upswing, from 1950 to 1970, growth continued but at a weakening pace. After 1970, however, New IT employment stagnated; after 1975 it fell.

Figure 12.2 disaggregates the picture for major sectors. With the exception of copying equipment, all showed notable employment growth during the third Kondratieff depression. In the upswing of the fourth, their growth paths diverged. Scientific and industrial instruments (MLH 354), radio and electronic components (MLH 364) and radio, radar and other electronic capital goods (MLH 367) grew most rapidly; photographic and document copying equipment (MLH 351) and electronic computers (MLH 366) did particularly badly. Internationally, as in the UK, the fastest-growing sectors were initially triggered by military demand (Freeman *et al.* 1965, OECD 1968, Tilton 1971, Braun & Macdonald 1982). In the UK, from about 1970 several important sectors began to register sharp job losses; only radar, radio and electronic goods (the most defence-based) appeared immune.

The question then is how employment change has related to output change. This is even more difficult to measure, but we have tried to do so for selected New IT industries for the period 1948–81. Figure 12.3 strongly suggests that the relationship is similar to that reported by Freeman *et al.* (1982) for all British manufacturing, but that the inflexion points come later in time in the case of the IT industries: the phase of expanding output and employment lasted until 1970 (as against 1963) and the succeeding period of jobless growth ended only in 1974 (as against 1970). The most disquieting feature is that a phase of declining employment *and* output began in 1980 – only one year later than for all manufacturing. Table 12.1 suggests that employment, which grew relatively modestly at 3.7% from 1948 to 1970, was static from 1970 to 1975 and thence declined by no less than 4.2% per year; output growth progressively slowed from one period to the next. Other figures, based on the new 1980 Standard Industrial Classification, suggest that the fall may have slowed after 1981 but that it was still continuing down to

Table 12.1 New IT industries, UK: average annual growth rates, 1948–81

	1949–70	1970–5	1975–81
output	7.2	3.4	1.7
employment	3.7	0.0	−4.2

Sources: Census of Production; *Economic Trends*.

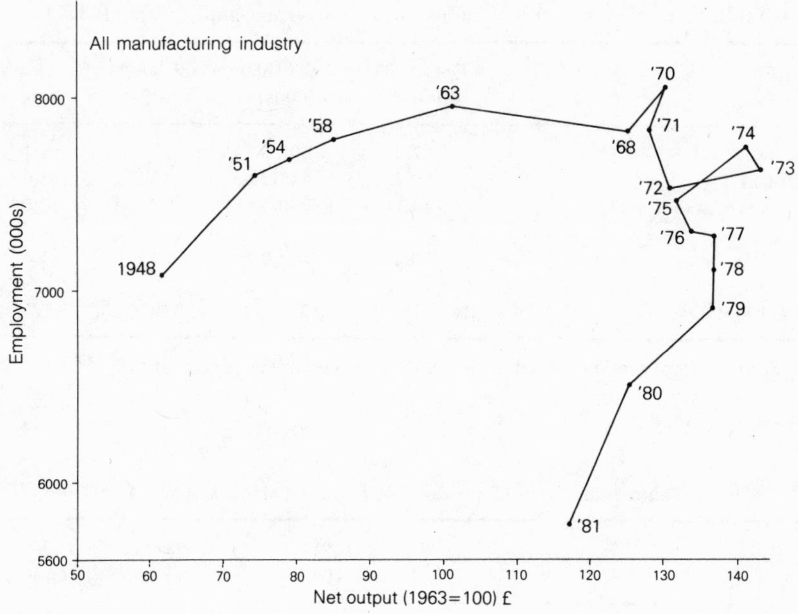

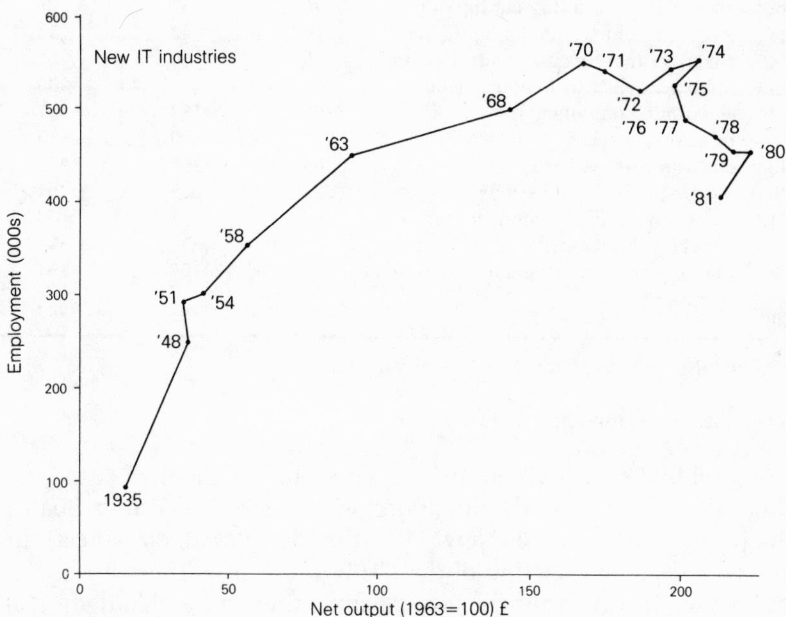

Fig. 12.3 Employment and net output, all manufacturing and New IT industries, UK, 1948–81, 1935–81

Table 12.2 New and Old IT industries and services: employment, 1981–4.

	Dec. 1981 (000s)	Dec. 1984 (000s)	Change 1981–4 (000s)	(%)
New IT	487.4	502.4	15.0	3.1
Old IT	457.7	441.2	−16.5	−3.6
IT services	744.1	770.4	26.3	3.5
total IT	1689.2	1714.0	24.8	1.5
total employment	21 117	20 771	−346[a]	−1.7

Source: *Employment Gazette*. SIC 1980. Some estimates necessary.

Note: [a] Approximate.

Table 12.3 New IT products: UK trade balances, 1982–4.

		1982 (£m)	1983 (£m)	1984 (£m)
3301	office machinery	−49	−77	−78
3302	electronic data processing equipment	−553	−872	−972
3433	alarms and signalling equipment	40	47	36[a]
3441	telegraph and telephone equipment	19	−43	−51
3442	electrical instruments, control systems	−16	10	61
3443	radio and electrical capital goods	252	317	283
3444	electronic components	−96	−132	−211
3452	records and tapes	1	−30	−29[a]
3453	electronic active components	−85	−286	−468
3454	electronic consumer goods	−921	−1055	−751
3710	measuring and precision instruments	126	87	64
3732	optical instruments	5	−15	−24
3733	photographic and cinematic equipment	−51	−157	−191
total		−1329	−2207	−2331

Source: *Business Monitor*, MQ10; and various.

Note: [a] Estimated from first three quarters.

1984 (Table 12.2). This picture is consistent with that based on other sources such as the Engineering Industries Training Board, which for 8 of our 13 New IT industries (1980 classification) showed a 1978–83 decline of over 16% (EITB 1984).

The conclusion appears inescapable: this more detailed, and longer-term, analysis confirms that the decline in British New IT employment after 1970 was not a function of general secular changes in productivity; it was a consequence of the deteriorating

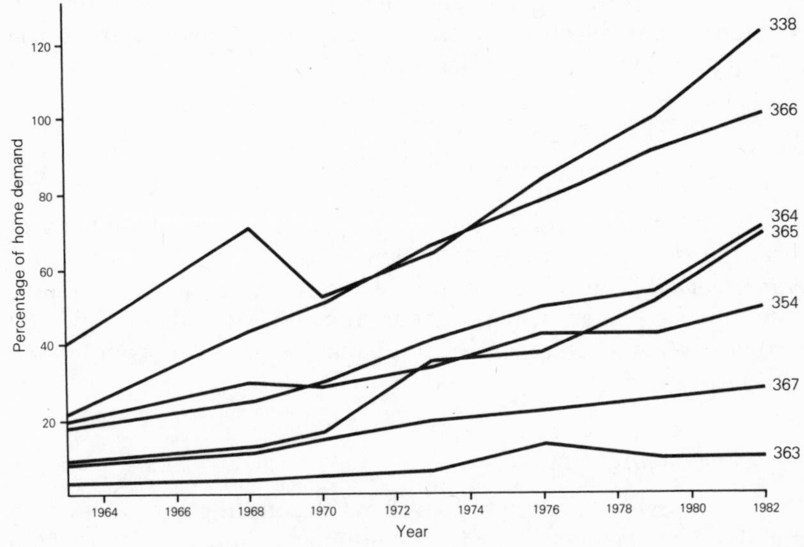

Fig. 12.4 Ratios of imports to home demand, New IT industries, UK, 1963–82

competitive position of British firms, associated with poor innovative performance.

Confirmation of this judgement comes from international trade figures. Though exports of New IT have risen (NEDO 1984), imports have risen even faster, causing progressive deterioration in the trade balance. Figure 12.4 shows this process over the period 1963–82, both for all New IT and for individual sectors, based on 1968 SIC categories; there is some variation from one sector to another, but the general trend is unmistakable. More recently (Table 12.3), figures based on the new 1980 categories show that the trend has continued; almost incredibly, the trade balance in New IT is if anything worse than that in all manufactured goods.

These findings can be confirmed by those of other studies. The analysis of NEDO (1984) shows that the electronics trade deficit has quadrupled between 1979 and 1983 to almost half of the combined deficit of all European countries. The core new IT industries had a trade deficit of £800 million in 1983, rising to £2.1 billion in 1984 and probably £3 billion in 1985 (ibid.). If software were included, the figures would be even worse. One forecast, by Cambridge Econometrics, suggested that on mid-1980s trends, despite a relatively strong export performance, imports would continue to grow rapidly and that the UK's trade deficit on electronics and electrical engineering products would treble by 1993 (Cambridge Econometrics 1985).

The anatomy of decline

To understand how this remarkable reversal happened we need to chart the progress of individual New IT sectors in the post-World War II period. The sources of decline are to be found in the fourth Kondratieff upswing, when – not so dramatically as later appeared – these sectors were already displaying critical competitive weakness.

(a) Office equipment

This sector, which includes typewriters, accounting and calculating machinery, document-copying equipment and similar office machinery, was marked by the rapid application of electronics from the late 1960s; even by 1968, for instance, about 25% of accounting and calculating machinery was electronic (Hays *et al.* 1969).

The British industry was singularly ill equipped to enter this competitive world, for it had long been losing ground. The typewriter, an archetypal third Kondratieff industry, provides an apt example. Since the start of the earlier long wave, its production had been dominated by the USA and Germany (Prais 1981); but, from 1970 onwards, Japanese producers came to supplant them in European and world markets. The UK, having obtained a foothold in the production of typewriters before World War II, persistently lost ground from the 1950s; by 1974 little remained of the industry (ibid.). In the early 1970s there were only three British plants producing typewriters, all owned by foreign companies; most of their production was for export, though most typewriters sold in the UK were imported (ibid.). Since the mid-1970s no typewriters suitable for office use were produced in the UK, though production of small lightweight portables appears to have continued. All production seems to have ceased by the early 1980s, though in the mid-1980s one major Japanese company announced plans to establish a plant in the UK to manufacture electronic typewriters.

In the entire sector, studies have suggested that British firms had a much lower level of R&D than their competitors, especially American firms (Hays *et al.* 1969), and that subsidiaries of foreign firms accounted for a large part of domestic output (Commission of the European Communities 1975). Thus, though output and employment rose down to 1970, the UK already accounted for a small and declining share of world markets.

(b) Consumer electronics

Radio production in the UK had boomed during the third Kondratieff depression of the 1930s, providing a classic case of a leading-edge industry that contradicted the general economic trends. After retooling for wartime needs, production of radio sets regained prewar levels by 1946 and reached a peak of about 2 million sets in 1947 and 1951, after which output declined.

Television production was restarted in 1946 after the government had decided to retain the 405-line system. Consumer expenditure on television overtook that of radios and radiograms in 1951. By 1955 every other British household had a television set: a figure behind the USA, where there was already the equivalent of almost one to every household, but well ahead of most other countries (Wilson 1964). High fidelity reproduction and other improvements to sound recording and reproduction, where British innovation played an important rôle, also contributed to the rapid growth of consumer electronics industries (ibid.). The BBC began engineering tests on colour television in 1955, though by then American firms such as RCA were well in the lead. One early analyst has suggested that the decision to adhere to the prewar 405-line system contributed to a technological lag, resulting in a loss of export markets as British standard equipment could only be sold abroad after costly modifications (ibid.).

The structure of the radio and television receiver producing sector was quite diverse. At the end of the 1950s there were about 40 assemblers of television sets and about 100 firms making radio sets and amplifiers on a significant scale (ibid.). The firms varied greatly in size, with about 10 very large firms accounting for three-quarters of production, but sitting side-by-side with a large number of quite small firms.

Consumer electronic industries remained the single most important electronics sector throughout the 1950s. From the end of that decade, however, they began a relative decline; and from the early 1960s, the capital goods electronic industries took the lead (Fig. 12.5). This shift occurred in most advanced capitalist economies, although the timing and pattern varied. The reasons are not hard to find. A spate of initial innovations (Ch. 9) coincided with strong growth of demand for such devices as computers, radar, navigational aids, electronic measuring, detecting and control equipment and

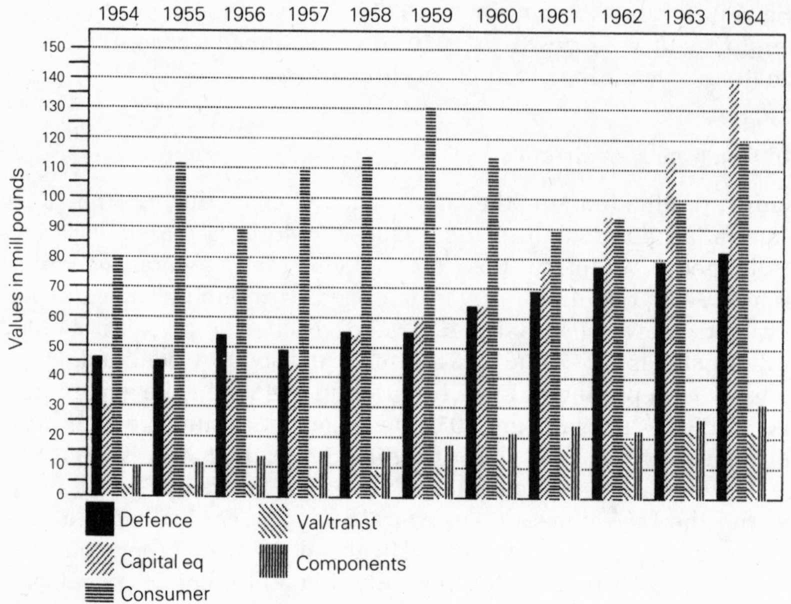

Fig. 12.5 Sectoral composition of output, electronics industries, UK, 1954–64

industrial control equipment. Overall, the electronic capital goods industries' share of total electronics output in the UK increased from 17.8% in 1954 to 35.1% in 1964 (NEDO 1967). The rapid growth of output at this time is summarized in Table 12.4.

(c) Computers

Here, though in 1946 the technological lead had clearly gone to the USA, British researchers were very close behind. Development work took place at London, Manchester and Cambridge universities, at Elliott Brothers research laboratories (with Admiralty support), and at the National Physical Laboratory (NPL), which co-operated with English Electric on the development of the ACE computer. In 1949 Cambridge produced EDSAC; Manchester University, in collaboration with Ferranti, produced MADM in 1950. Ten of these machines were ordered by the National Research Development Corporation (NRDC) from Ferranti.

Over the following decade the NRCD supported several other projects and organized a patent pool. Yet in the first decade its total financial support for development amounted to less than a million pounds (Freeman *et al.* 1965). At this time, British electronics

Table 12.4 Electronic capital goods, estimated deliveries, UK, 1958–65.

	1958 (£m)	1961 (£m)	1965 (£m)
industrial electronic control equipment	2.2	11.9	58.1
electronic computers	5.3	10.9	34.6
nucleonic instruments	n.a.	n.a.	4.9
electronic measuring instruments	8.0	11.7	26.6
broadcasting transmitters	3.5	5.5	6.1
radar and navigational aids	15.5	14.1	71.1
radio communications equipment	9.9	13.4	36.1
other electronic equipment	5.2	9.9	34.6
total	49.7	77.3	268.4

Source: Ministry of Technology data, cited in NEDO (1967).

Note: The series has been subject to many breaks in the period covered. There was an increase in the numbers of firms covered by the survey over successive years; this and other breaks tend to exaggerate annual increases in some sectors.

companies seem to have been remarkably confident about their competitiveness: at a meeting at NRCD in 1949 they pronounced that 'they were individually in positions to tackle the problems of an electronic computer development project as well as, for example, the International Business Machines Corporation in the United States' (Crawley 1957). The Lyons company developed the LEO computer in collaboration with Cambridge University; although an early model was used for bakery costing work from 1951, the first delivery to an outside customer was not until 1958. The British Tabulating Machine Company (BTM), an established supplier of punched card and other such machines developed by IBM, also designed and developed a small machine in collaboration with Birkbeck College in London. This machine, the Hollerith 'HEC', was first delivered in 1955; in 1956 BTM set up a joint company with GEC to develop a new set of computers, resulting in the transistorized ICT 1301 of 1962.

In 1959 BTM merged with Power Samas to form ICT; in 1962 ICT took over the EMI computer division; in 1963 it absorbed the Ferranti computer division. In 1964 English Electric (together with its subsidiary Marconi) merged its computer interests with LEO and with Elliott Automation in 1967. By this series of mergers, the 10 firms involved in developing and manufacturing computers in 1950 had been reduced to 3 by 1964 (excluding IBM and Honeywell of the USA) and this was reduced to one on the formation of International Computers Limited (ICL) in 1968 (Fig. 12.6). ICL

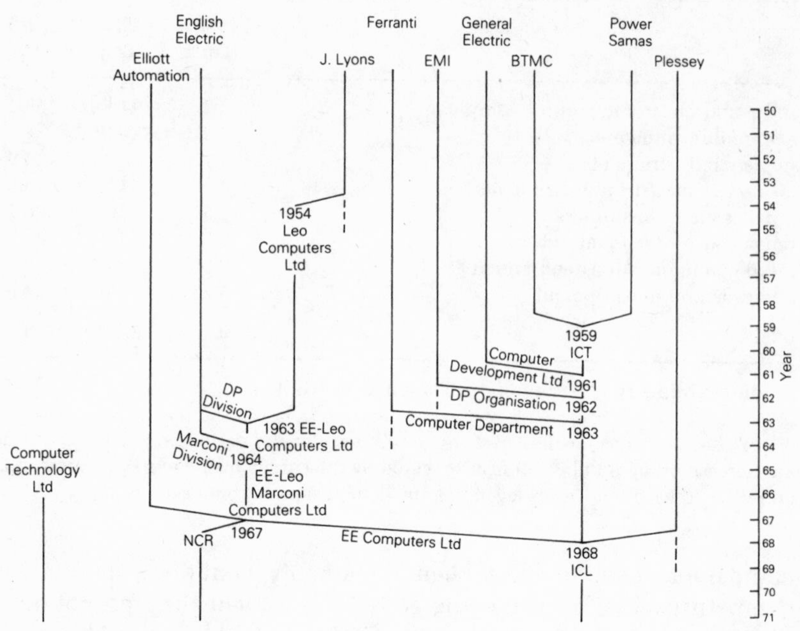

Fig. 12.6 Computer manufacturers, UK, 1950–70
Source: Sciberras (1977).

became the fourth largest computer firm in the world; its resources gave it potential competence in every major area of computer and components activity. In 1968 ICL had an estimated 45% of the British computer market, against IBM's 40% (Freeman *et al.* 1965, Harman 1971, Stoneman 1976).

Thus in 1960, despite relatively low levels of government support and fewer and slower government orders than in the USA, a number of British firms had made successful first-generation valve machines in the UK; but few models had sales in excess of 50 (Freeman *et al.* 1965). The lack of government demand was exacerbated by the relatively low levels of per capita computer use compared to Germany and especially the USA. No computer firm made any profit from them before the 1960s, so there was little incentive for swarming (ibid., Harman 1971). Despite its pioneering work, the industry in the UK lagged behind that in the USA in developing second-generation transistorized machines in the early 1960s; and the new British computers of those years contained significant American technology import content (Freeman *et al.* 1965).

British firms appeared to enjoy much more success in scientific and other specialized markets than in commercial EDP markets where customers required higher standards of software, training advice and support services (ibid.). This is borne out by the parallel industrial control equipment sector of the electronic capital goods market. The leading British firms in this field (Elliott Automation, Ferranti and English Electric) accounted for about 13% of world industrial control computer installations in the early 1960s (ibid.). At this time the UK was estimated to lead other European countries in the number of control computer installations and in the number of machine tools with electronic control systems, though the USA already had about ten times as many installations of both as the UK. Ferranti pioneered the development of electronically controlled machine tools (the NPL also made important contributions) and had the largest share of the market in the UK for the most sophisticated types. It also developed an associated inspection machine which obtained large export orders including to the USA (ibid.).

By the mid-1960s it was evident that the UK's performance and prospects in computers were much stronger than in more traditional office machinery industries; the British computer industry was the only one in Europe with prospects for a viable, independent future (Hays *et al.* 1969, Harman 1971). The UK was the only country other than the USA with indigenous producers designing and producing their own computer equipment, apart from a few firms making scientific or specialized equipment. European computer production outside the UK was entirely dominated by American firms. At this time, only in the UK and Japan was the American giant IBM's share of computer installations less than half; overall, it had about the same proportion of installations and sales outside the USA as within it, about two-thirds (Freeman *et al.* 1965). Yet, even down to 1965, American computers accounted for over half the total sales value of electronic data-processing equipment and about half the total number of installations in the UK.

(d) Radio capital equipment

In one specialized field, radio broadcasting equipment, two firms dominated the British home market at this time and accounted for most of the exports: Marconi, a subsidiary of English Electric since

1946; and STC, a subsidiary of the American giant ITT (Wilson 1964, Freeman *et al.* 1965). Marconi continued its traditional strong performance in this sector throughout the 1950s and early 1960s and accounted for more than half of British exports.

Another sector, specialist radio communications, also grew rapidly in this period, partly because of military demand. The USA had a dominant position with three-quarters of world production and almost the same share of world exports. In addition a major American firm, ITT, had subsidiaries engaged in production and export sales in many European and other countries, including the UK (STC). Despite this, some British firms (Plessey, Racal and Redifon) had secured a base in both domestic and exports markets. Racal and Redifon specialized in high-frequency radio communications equipment and both exported more than half their output in the early 1960s (Freeman *et al.* 1965).

In radar and navigational aids equipment, the USA – with massive programmes for early warning systems, missile and satellite undertakings – proved again the dominant market and producer. But some British firms, principally Decca and Marconi, were very successful in marine and airborne radar and navigation systems, with exports accounting for very high shares of output. It was estimated in the early 1960s that British firms supplied the majority of radar systems installed in all ships in civilian airports worldwide (ibid.).

In the navigational aids field, British firms such as Decca, Marconi, Smiths, Elliotts had a strong technological position and exports were relatively high. Indeed, in the mid-1960s, the UK seized a much higher share of total leading-country exports in these capital informational equipment sectors such as radio transmitters (17.4%), industrial electronic control equipment (35.8%) and radar and navigational aids (29%) than in computers where its share was only 4.8% (ibid).

The relative growth of electronic capital goods industries contributed to a significant increase in overall electronics exports throughout the postwar boom years, as Figure 12.7 illustrates. But from the mid-1960s, the growth of imports began to quicken and this worked to reduce the positive trade balance (NEDO 1967, 1968).

(e) Active electronic components and semiconductors

In the early 1950s, production of active electronic components in the UK was more heavily concentrated than in the interwar years, but the relative position of the major companies had changed

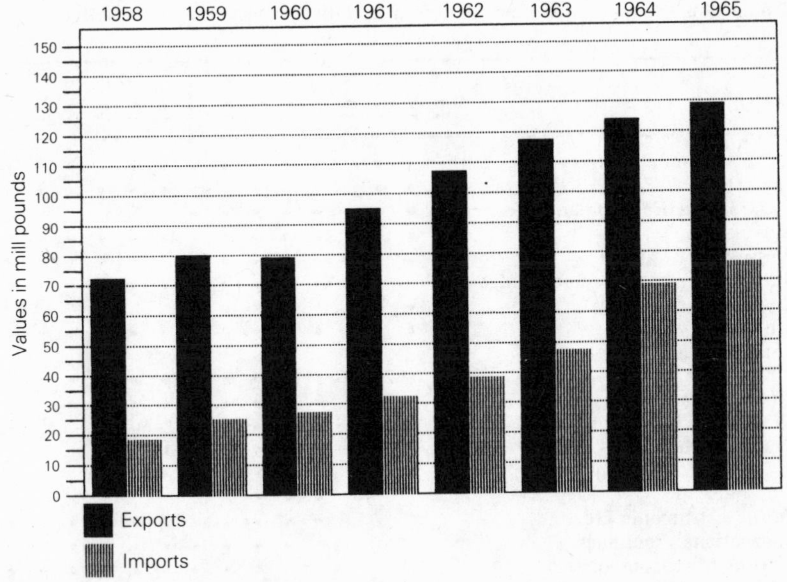

Fig. 12.7 Electronics exports and imports UK, 1958–65

Table 12.5 Major UK producers of valves and cathode-ray tubes, 1954 (percentage of total output).

	Valves	Tubes
Philips (Mullard)	58.5	37.1
AEI (Edison-Swan and BTH)	13.2	31.2
Standard (ITT)	13.3	4.5
EMI	8.0	16.8
GEC	3.7	4.8
Ferranti, EEC, Pye and others	3.3	5.6
total	100	100

Source: Allen (1970).

Note: The English Electric Valve Company and Cathodeon were also producing television tubes in the 1950s (Wilson 1964).

somewhat, as Table 12.5 illustrates. The largest producer was Mullard, now a subsidiary of the Dutch firm Philips. Together with the ITT subsidiary STC, these two overseas-based multinationals accounted for well over half the UK's output of valves and tubes, indicating a weak performance by British firms compared to the inter-war years.

Table 12.6 Semiconductor firms in Great Britain and years they were active in the industry, 1954–68.

Firm	1954	1955	1956	1957	1958	1959	1960	1961	1962	1963	1964	1965	1966	1967	1968
AEI Semiconductors	•	•	•	•	•	•	•	•	•	•	•	•	•	•	•
Ferranti	•	•	•	•	•	•	•	•	•	•	•	•	•	•	•
General Electric Company Ltd.	•	•	•	•	•	•	•	•	•						
Newmarket	•	•	•	•	•	•	•	•	•	•	•	•	•	•	•
Plessey	•	•	•	•	•	•	•	•	•	•	•	•	•	•	•
Standard Telephones and Cables	•	•	•	•	•	•	•	•	•	•	•	•	•	•	•
English Electric Valve		•	•	•	•	•	•	•							
Mullard		•	•	•	•	•	•	•	•	•	•	•	•	•	•
Westinghouse Brake English Electric			•	•	•	•	•	•	•	•	•	•	•	•	•
Joseph Lucas				•	•	•	•	•	•	•	•	•	•	•	•
Associated Transistors					•	•	•	•							
Brush Clevite					•	•	•	•	•	•	•	•			
MCP Electronics					•	•	•	•	•	•	•				
Stone-Platt					•	•	•	•							
Texas Instruments Ltd.					•	•	•	•	•	•	•	•	•	•	•
International Rectifier						•	•	•	•	•	•	•	•	•	•
Emihus Microcomponents							•	•	•	•	•	•	•	•	•
Microwave Semiconductor Devices									•						
Societa Generale Semiconduttori-UK									•	•	•	•	•	•	•
Semitron										•	•	•	•	•	•
Elliott Automation Microelectronics											•	•	•	•	•
Marconi Elliott Microelectronics												•	•	•	•
Transitor												•	•	•	•
Thorn															•

Source: Sciberras (1977).

At this time, the new transistor technologies began to be diffused from the USA, more rapidly to the UK than to other European countries (OECD 1968, Tilton 1971). In the early and mid-1950s this occurred through licensing agreements to UK producers, but thence – following the lead of Texas Instruments in 1958 – some of the major new American semiconductor firms established subsidiaries. Table 12.6 lists the major producing firms in the UK in the 1950s and 1960s and the years in which they were engaged in production. As elsewhere in Europe, and quite unlike the USA, few new firms emerged in the new technologies (OECD 1968, Tilton 1971, Sciberras 1977, Freeman *et al.* 1965, 1982).

By 1968 the established major receiving tube firms (Mullard, STC and AEI) had about 32% of the British semiconductor market; the 'new' firms (Ferranti, Westinghouse Brake/English Electric, Marconi Elliott Microelectronics and a few others) accounted for only 16%; the new foreign subsidiaries (Texas Instruments, Societa Generale Semiconduttori (SGS), International Rectifier and others) held no

Table 12.7 UK semiconductor industry: leading firms, 1962 and 1973.

1962	%	1973	%
ASM	49	Texas Instruments	18
Texas Instruments	13	Mullard	17
Ferranti	10	Motorola	14
AEI	7	ITT	14
Westinghouse Brake	5	Ferranti	5
STC	2	Plessey	4

Source: Sciberras (1977).

less than 44% of the market; importers such as Motorola accounted for 8% (Tilton 1971). By the early 1970s the top producers were all subsidiaries of American or European multinationals (Table 12.7). Although the UK was a major European location for many American firms, the value of semiconductor imports exceeded exports throughout the 1960s. Clearly, then, the radical restructuring of the industry brought by the new semiconductor technologies spelt serious trouble for the British industry. The major British firms involved in semiconductor production, such as Ferranti and Plessey (each with less than 5% of the domestic market by 1972–3) proved unable to compete with the international giants in the large-volume standard component markets, and were confined to specialist markets such as the production of short-run components for military uses (Sciberras, 1977).

Conclusion

So, even in the upswing years of the fourth Kondratieff, the UK was already falling out of the world technological race. Even from the start – the first digital computer in 1946, the transistor in 1947 – it was clear that a technology gap had opened between the USA and all its competitors. Yet the gap was less evident for the UK than for others: for a time, British companies remained innovative in fields like computers and specialist military markets, where relatively high defence spending provided support; and licence agreements, plus the establishment of American subsidiaries, helped rapid diffusion of American technologies (OECD 1968, Sciberras 1977, Braun & Macdonald 1982). But by the mid-1960s it was already clear that leadership had passed to the USA, as evidenced by the lack of new firms swarming in critical innovative areas and the domination of American subsidiaries in many sectors. The question must now be why.

Towards explanation: the rate of innovation

The starting point of the explanation must be this: that British competitive decline – in IT as in other high-technology industries, indeed in industrial production generally – was no sudden phenomenon of the 1970s and 1980s. On the contrary: though first widely noticed then, it was merely the acceleration of a phenomenon that had begun in the late 19th century, and was evident not only in the successive high technology sectors examined in this work, but throughout the industrial system. This was crucial: a country like the UK, with low general innovative potential and performance, will provide poor ground for the growth of the high-technology sectors, because home demand for their products – vital in their early stages – will be weak.

One obvious measure of this is the UK's poor level of technological innovation. The myth has it that the UK has been strong in invention and initial innovation, weak in commercial follow through. This may have been true once; it has not been evident in the fourth Kondratieff, during which the UK's overall share of inventions, as measured for instance by patent registrations, has declined dramatically (Table 12.8). In the information technology industries, Chapter 7 has shown that as early as the start of the third Kondratieff the UK had been losing the lead in basic innovations to Germany and the USA, both in the new electric sector and in the mechanical and electromechanical office technology sector. True, in early electronics it had been strongly competitive down to World War II and just after; then, as the first half of this chapter has chronicled, the story was one of rapid decline followed by collapse. Again, comparative data tell the story (Tables 12.9 & 12.10). Over the whole 1963–81 period, of total IT–related patents

Table 12.8 US patents granted to inventors from selected countries, 1966–82.

country	1966	1970	1974	1978	1982
total	68 405	64 429	76 278	66 102	57 889
USA	54 634	47 077	50 649	41 254	33 869
foreign	13 771	17 352	25 629	24 848	23 993
of which:					
West Germany	3981	4435	6153	5850	5409
Japan	1122	2625	5888	6911	8149
UK	2674	2954	3145	2722	2134
France	1435	1731	2566	2119	1975

Source: National Science Board (1983).

Table 12.9 US patents granted in IT-related fields, 1963–81[a].

	USA	West Germany	Japan	UK	Total[b]
electrical equipment except communications equipment	96 556	8462	8416	5317	133 272
communications equipment and electronic components	101 914	7850	13 013	5976	143 156
professional and scientific instruments	103 333	11 145	13 353	5478	149 425
aircraft and parts[c]	16 931	2748	2637	1845	27 201
total	865 124	91 359	77 450	51 138	1 234 643

Source: National Science Board (1983).
Notes:
[a] This source lists the total number of patents granted to inventors in the USA and 15 leading countries over this period in terms of 15 categories. The four fields chosen are those most relevant to IT broadly defined, but some inevitably include non-IT-related patents.
[b] Totals refer to all product fields patents for the 15 leading patenting countries.
[c] A major user and adaptor of electronic-based IT products and systems.

Table 12.10 Major country shares of selected foreign patenting in USA, 1981

Product field	West Germany	Japan	UK	others[a]
electrical equipment except communications	21	37	8	34
communications equipment and electronic components	18	44	9	29
professional and scientific instruments	22	43	8	27
aircraft and parts	28	42	10	21

Source: National Science Board (1983).
Note:
[a] Other Foreign countries' shares.

taken out in the USA, the British share was a mere 4%; in 1981, the British share of all foreign patents was as low as 8–10%, far behind its two major competitors.

Technological transfer might have filled this gap, and the evidence is that, at least from the USA, this was very rapid, more so than for other advanced industrial countries. Yet the UK does not

Table 12.11 Estimated civilian R&D ratios for selected countries, 1961–78[a].

	Percentage of GNP				
	France	West Germany	Japan	USA	UK
1961	0.97	(1.10)	1.37	1.20	1.48
1964	1.34	1.38	1.47	1.27	1.49
1967	1.50	1.70	1.49	1.48	1.65
1970	1.47	1.96	1.77	1.50	1.66[b]
1973	1.30	2.02	1.86	1.44	1.48[b]
1976	1.36	2.10	1.89	1.50	1.38[b]
1978	1.36	2.13	1.90	1.54	1.49

Source: National Science Board (1983).
Notes:
[a] National expenditures for R&D excluding government funds for defence and space purposes; includes associated capital expenditures except for USA where data are not available.
[b] Data are for year previous to that of the other countries.
Brackets indicate estimates.

Table 12.12 National expenditures for R&D, 1969 and 1979.

	France ($b)	West Germany ($b)	Japan ($b)	USA ($b)	UK ($b)
1969	2.7	3.1	3.0	25.6	2.5
1979	11.0	20.9	19.3	55.0	7.1

Source: National Science Board (1983).
Note: Gross expenditures for performance of R&D including associated capital expenditure except for USA where data were not available.

appear to have benefited, probably because much of the transfer came via American-owned multinationals and thus did not promote a virtuous circle of indigenous innovations, firm growth and industrial swarming.

The R&D factor

Since high-technology industry is essentially research and development embodied in products, the most obvious explanation of this weakening performance would seem to lie in the resources devoted to R&D and to associated education and training activities. The empirical evidence, which is also fraught by many difficulties of international cross-comparison, suggests that indeed the UK has increasingly lagged behind its competitors both in absolute and relative terms. Tables 12.11 and 12.12 demonstrate this for total civilian R&D: during the period 1961–78 the UK fell from top to fourth place in the proportion of GDP devoted to R&D (Table 12.11); between 1969 and 1979 its absolute level of R&D spending was the lowest of any of the major industrial countries, and the gap was growing (Table 12.12).

More recent evidence, assembled by OECD, is summarized in Table 12.13. It confirms that the UK's absolute level of R&D was lower than that of any of its major competitors; that its spending relative to GDP was lower than any other country save France; and that the gap was widening. It also shows that the R&D performance of the Business Enterprise Sector (BERD) was particularly poor, especially in the contribution of business itself to research and

Table 12.13 Selected R&D indicators, major OECD countries, *c.*1983.

	France	West Germany	Japan	USA	UK
Gross domestic expenditure on R&D[a]					
level in 1983[b]	13.1	18.1	31.5	88.4	12.5
annual average growth 1980–3[c]	4.3	2.1	9.9	4.5	(1.0)
GERD as percentage of GDP 1983[d]	2.1	2.6	2.7	2.7	2.3
Percentage GERD industry-financed[e]	42.0	58.1	64.0	48.8	42.0
Gross Expenditure on R&D in the business enterprise sector (BERD)					
BERD level in 1983[b]	7.4	12.6	19.7	62.9	7.7
estimated annual average percentage growth 1981–3	4.1	2.7	(9.3)	5.2	(0.5)
BERD as percentage of GERD 1983[d]	56.6	69.8	62.5	71.2	61.3
percentage BERD industry-financed[f]	68.2	81.7	97.9	68.4	61.3

Source: Authors' calculations of OECD data quoted in annual Review of Government Funded R&D 1985.

Notes:

[a] Gross domestic expenditure on R&D (GERD).

[b] In billions US dollars, current prices, purchasing power parities; Japan data estimated from 1982 level and growth rates.

[c] At percentage of fixed prices; calculated from arithmetic means of the figure for the best three years given except for UK where only two years' data are available (i.e. 2.9% in 1981 and −0.9% in 1983).

[d] Japan figure estimated from 1982 actual and growth rate.

[e] 1983 data, except estimate from 1981 and 1982 data for Japan.

[f] 1981.

Table 12.14 Changes in UK intramural industrial R&D expenditure, 1975–83 (£m, 1975 prices).

	1975	1978	1981	1983
all product groups	1340	1566	1661	1564
all products of manufacturing	1293	1512	1555	1465
chemicals and allied products	254	284	277	279
mechanical engineering	103	118	111	99
electronics	279	442	511	529
other electrical engineering	73	69	53	45
motor vehicles	88	88	80	91
aerospace	291	285	337	272
other manufactured products	213	226	185	151

Source: British Business, 18 January 1985.

development, and that the gap *vis-á-vis* the other main industrial nations was growing at a positively alarming rate. This is confirmed by a study reported early in 1985, which indicates that R&D spending by British industry actually fell in real terms between 1981 and 1983. Perhaps the only comforting feature of this analysis is that R&D spending on electronics showed a notable increase, both from 1981 to 1983 and over the longer period from 1975 to 1983 (Table 12.14).

The evidence could be multiplied from numerous sources, but that is unnecessary: it overwhelmingly suggests that in the 1980s the UK was the only major industrial country to record an actual fall in the proportion of GDP devoted to R&D, whereas industrial R&D spending was increasing at a negligible rate in comparison with competitor countries (Marsh 1985).

The military bias

This, however, is not the end of the story. A high and growing percentage of the UK's effort went on military applications rather than fundamental science or industrial technology. In 1983 the share of British GDP devoted to civilian R&D was the lowest of any major industrial country; the percentage spent on military applic-ations was greater than any except the USA (Table 12.15). And expenditure plans suggest that down to 1987–8 this discrepancy will, if anything, increase. This explains both the high proportion of British R&D that is state financed and the concentration of that state effort on the Ministry of Defence (Table 12.16): in 1981–2 the MoD share of state R&D spending was already close to 50% and was expected to rise to 55% by 1987–8, whereas the DTI share was programmed to fall from 8% to just over 6%. One interesting implication, clear from Table 12.17, is that the effort is heavily concentrated on electronic and aerospace technologies; nearly 90% of state spending on electrical and electronic engineering came from the MoD. This in turn helps explain the fact that in 1983 no less than one-third of all industrial R&D spending went on electronics (Table 12.18). Much of this military expenditure has gone to large-scale, prestigious and exotic projects (Peck 1968; Freeman 1978; Pavitt 1980), where perversely – because the USA invested in precisely these technologies on a much larger scale – the commercial spin-off proved negligible (Freeman 1978).

Despite this concentration, in fact, British electronics R&D already lagged far behind that of the USA by the late 1950s, whether measured absolutely (less than one-seventh), per research worker (one-third) or as a percentage of net output (less than half)

Table 12.15 Estimated civilian and military R&D, major OECD countries, 1983

	percentage of GDP				
	France	West Germany	Japan	USA	UK
total civil R&D[a]	1.7	2.5	2.5	2.0	1.6
state defence R&D expenditure[a]	0.46	0.11	0.01	0.76	0.65

Source: Cabinet Office (1985).

Note:
[a] Because gross and defence R&D expenditures are calculated in different ways and for other reasons, these two quantities are not strictly comparable. The above figures are simply a guide to the proportions of national wealth spent on civil and military R&D.

Table 12.16 Shares of UK state-funded R&D by major departmental grouping.

	Percentage shares of total		
	1981–2	1984–5	1986–7 (plan)
total civil departments	24.4	23.2	21.9
of which: DTI[a]	8.1	8.6	6.4
total research councils/UGC	25.8	25.4	25.1
total civil	50.2	48.7	45.0
total defence	49.8	51.3	55.0

Source: Cabinet Office (1985).
Note:
[a] Department of Trade and Industry.

Table 12.17 State R&D most related to specific industrial sectors, 1983–4[a]

	MOD[b] (£m)	DTI[c] (£m)	SERC[d] (£m)	Others (£m)	Total (£m)
extractive industries etc.	—	2.8	2.4	0.7	5.9
products of manufacturing industry[e]	—	8.4	3.5	—	11.9
chemical industry	25.0	3.8	6.1	1.3	36.2
metal goods	—	—	—	—	—
mechanical engineering	221.0	22.0	9.7	3.1	255.8
office machinery etc[f]	47.0	28.1	0.1	—	75.2
electrical and electronic engineering	408.0	43.1	1.6	2.1	454.8
aerospace equipment manufacture and repair	628.0	107.3	0.2	—	735.5
other manufacturing industry	103.0	22.0	3.3	16.7	145.0
construction	—	0.3	2.7	8.3	11.2
utilities, services etc.[g]	—	1.7	0.6	182.1	185.0
other	10.0	66.1[h]	20.4	33.5	130.0
total[i]	1442.0	305.6	50.5	248.3	2046.4

Source: Cabinet Office (1985).

Notes:

[a] This is defined as government R&D spending to improve technology or for the support of procurement decisions (see section 4.1 of source).

[b] Ministry of Defence.

[c] Department of Trade and Industry.

[d] Science and Engineering Research Council.

[e] Mainly iron and steel, non-ferrous metals, bricks, pottery and china.

[f] Includes electronic data processing equipment where MOD funded R&D was £47.0 million and DTI £25.6 million.

[g] Includes £10.0 million on postal services and telecommunications.

[h] The main DTI expenditures here were: space £60.4 million; process industry £2.1 million; computer software and hardware £1.7 million.

[i] Only spending which could be allocated to industrial sectors is included in this table. The totals may not be entirely consistent.

(Freeman 1962). This was partly the result of vast American state funding, which gave companies a massive lead in all stages of the research–development–production continuum, allowing them progressively to improve products and develop mass-production applications (Freeman *et al.* 1965, OECD 1968, 1969, 1970, Tilton 1971, Sciberras 1977, Braun & Macdonald 1982, Freeman *et al.* 1982, Dummer 1983). British firms could not compete in mass markets, or indeed in basic military applications; they retreated into specialized niches, thus inhibiting further innovation in

Table 12.18 Industrially performed R&D, UK, by major product groups (£m, constant 1975 prices).

	1966	1972	1978	1981	1983
all product groups	1504	1417	1566	1661	1564
all manufactured products	1478	1376	1511	1554	1465
including:					
electronics	288	301	442	511	528
other electrical engineering	94	68	69	53	45
aerospace	330	366	285	337	272
mechanical engineering	155	94	118	111	99

Source: CSO (1985).

consumer electronics fields such as computers, television and telecommunications.

There is an obvious paradox here: American military R&D produced massive market spin-off, British did not; the Japanese, from a starting point well behind the UK, triumphed without any military research base. One explanation is that the Americans used military spending as a covert industrial development policy – for instance, by funding the development costs of new computers and by buying exclusively from American manufacturers – whereas the British, still far more wedded to traditional *laissez-faire* policies, did not. Thus, ironically, ICL found its products rejected by the British Ministry of Defence, which bought them, under protest, only for payroll and other administrative applications: a factor that contributed to its crisis in 1982. This may be because key American personnel adopted a conscious developmental approach, particularly in the mid-1950s when the ICBM and space programmes were launched; it may also reflect the relative freedom of new firms to enter into competition following the successful antitrust action against Bell Laboratories.

The British weakness is especially clear in computers, where – even after the amalgamations of the 1960s – ICL's R&D resources were dwarfed by the American and Japanese giants. As a result, original research progressively declined. By 1985, an authoritative source could comment:

> Original research into IT areas in the UK is at an all time low, even accounting for government-supported projects. There is little investigation of original storage and processing materials for components, little work on complex software systems which will change the methods of commercial system building, hardly

> anything of note in the commercial field on processor design and almost no significant work on peripherals developments. That is not to degrade the efforts of researchers, particularly academic researchers in the UK. But none of them can honestly claim that any research programme currently under way or any amalgam of programmes will have the type of effect that the Manchester University work had in the 1960s or the Cambridge University work in the same period. (Anon.1985).

The Alvey programme, designed to compete with the Japanese fifth-generation computer project, seemed to have been similarly distorted towards military applications and small firms claimed that they had been effectively excluded by the counterpart funding requirements. By then the British industry no longer represented an effective self-generating industrial complex:

> The UK IT industry is dead. To earn the title of an IT industry a nation must have a single integrated industry that can sustain a broad range of developments and products, supported by original research in key sectors and with a significant portion of world markets. The UK now has none of these attributes. . . . As for having an integrated industry with a broad range of developments and products – the litmus test of a mature IT industry – ICL's position shows how far the UK has slipped. This flagship of the UK industry is dependent on Japanese components to get the performance for its range of mainframes. It is dependent on a Canadian telecoms supplier for its communications switch and a range of other non-UK suppliers for other vital components and sub-systems. (Ibid.)

To sum up, British R&D was too limited, too little state-supported in its crucial industrial applications, all too heavily state supported in military applications. These distortions were observable in the 1950s, when they contributed to the speed at which the UK's competitors – Germany, then Japan – overtook it in commercial high-technology applications (Shonfield 1958). They have, if anything, increased in recent years. In contrast, the USA, which spent less on defence as a proportion of national income, but used it to ensure political leadership of the west, had a highly effective industrial development policy, while claiming that it had none: it targeted certain technologies, funded them through the military budget, and provided a guaranteed market for them; it also protected them from outside competition, claiming military security as the reason.

This, it may be claimed, was only possible because the enormous absolute size of the American military budget facilitated this exercise in global monopoly. Persuasive critics have recently suggested that, whatever the success of this policy in the 1950s, in the 1980s it is imposing serious handicaps on the USA in its global competitive struggle for New IT supremacy (Stowsky 1986a, 1986b, Cohen & Zysman 1987). We return to these criticisms in our final summing up. What still remains, surprising, given the determination of British governments to follow this military–led route, is how monumentally unsuccessful they were from the start.

13
The flight from Electropolis: innovation on the move

As on an international stage, so on principal national stages, the fourth Kondratieff was an era of profound geographical shifts in the production of the New IT. Internationally, Chapters 10–11 have chronicled the relative decline of Europe, the continuing dominance of the USA, the challenge of Japan, and the rise of a few NICs. Nationally, within each of these countries, there was a general shift away from the older giant metropolis, toward newly emerging innovative centres: Berlin and London and New York lost, southern Germany and South East England and Silicon Valley emerged. The focus of this chapter is on these local shifts: on how, and why, they occurred.

In fact, the fourth Kondratieff history of the great metropolitan innovation centres proves to depend partly on general decentralization tendencies, partly also on features special to each. The fall of Berlin, the great Elektropolis, resulted uniquely from the shock of defeat and destruction. London's decline as a centre of electronic innovation was in contrast gradual, dependent partly on outward movement of firms encouraged by government policies of the 1950s and 1960s, partly on the progressive failure of British competitive power. New York's loss of position, again in contrast, was relative only: it arose from the rapid, even spectacular, rise of the new electronic complexes of Greater Boston and Silicon Valley. Though the second of these was truly a new phenomenon, the first – somewhat like the London case, but far more successful – represented the outward diffusion of a much older third Kondratieff innovative tradition.

The fall of Elektropolis

Berlin in 1945 – as already seen in Chapter 8 – was left physically shattered and effectively denuded of its industrial base; the subsequent years of the airlift and the city's isolation did nothing to repair that damage. Meanwhile, Germany's electrical and electronic industries were replaced elsewhere. By 1967 Berlin accounted for only 10% of electrical engineering employment in the Federal Republic against 24% in Bavaria or 20% in Baden-Württemberg, the two areas that had benefited most from Berlin's decline (Schulz-Hanssen 1970). There was worse to come: between 1960 and 1983 manufacturing employment in West Berlin fell by nearly 50%, from 305 000 to 157 000. By 1981 Berlin in fact had a deficit of employment in electrical engineering, particularly marked in skilled work and in higher-level managerial posts (Ewers 1984).

This should not be surprising. Berlin's very advantages had now for the most part turned into disadvantages. The close ties to the state agencies had been severed with the establishment of the Federal Republic and the effective end of all Berlin's capital city functions. The existing workforce was ageing and the new reservoirs of skill were created in the southern parts of the Federal Republic. The magic circle of contact between industry and scientific research, despite the continued prestige of the 'TU', had been broken.

This had more than a local significance; in losing much of the Berlin industrial complex, Germany lost a unique economic asset that could not simply be replaced elsewhere. Although its electrical industry steadily grew in output and employment, it could never again hope seriously to challenge American leadership in the emerging field of electronics. The historian of Siemens has written:

> Although not fully appreciated at that time, the history of electricity has shown repeatedly that a new shoot will stem from the common trunk and endeavour to develop into a sturdy branch with a life of its own. (Siemens 1957, 186)

Thus the telegraph, an innovation of the second Kondratieff intimately associated with the railways that were one of the central advances of that period, had produced a tradition of research that begat the new lighting and traction industries as well as the information-technology successors of the telegraph: the telephone and the radio. These, in turn, had helped trigger the third Kondratieff of the 1880s and 1890s. But, despite promising signs in the 1930s – an early lead in sound films, in telex and in military communications – at the onset of the fourth Kondratieff the chain was broken. Germany had lost much of its finest technical expertise through emigration in the Nazi years, and the spoils of war included direct access to what was left. It recovered strongly to make important advances in such fields as medical technology, but it could never again challenge the USA at the leading edge of the new technologies. This must have consequences for the wider economy.

Perhaps that would have been true anyway, as a comparison with London may suggest. By the end of the third Kondratieff there were clear indications that, having seized technological leadership from the UK and shared it with the USA, Germany was in the course of having to concede victory to its transatlantic rival. But the fall of Berlin was of incalculable consequence, the importance of which can never be estimated. Almost at one blow, the accumulated tradition of innovation – the close links between university and industrial research, the network of large and small firms, the accumulated research of the large organizations – was destroyed. The city could not replicate them, because it had ceased to be the capital of a major industrial nation, and because critical links between government, industry and research were thus broken. With this loss went some important part of the country's innovative power to compete. That is the tragic lesson of the fall of Elektropolis.

London's Western Corridor

In Chapter 12 we saw how, in the UK, progressively weaker growth of New IT employment finally gave way to decline at the start of the fourth Kondratieff downswing. Against this background it is possible to appreciate the long-term trends in the location of the New IT industries in the UK. The outstanding feature is their heavy concentration in the South-East. There is however a fairly sharp decline in the South-East's share from 74% in 1911 to 56% in 1951, after which it is fairly constant at 52–60% (Table 13.1). The only other region that appears at all significant is the North-West, which over a long period has recorded between 7% and 17% of New IT; the general trend appears to be clearly downward.

These figures, of course, refer to percentage shares. Table 13.2 shows for the New IT industries the changes in actual figures between 1971 and 1981. Nationally, as already seen, employment in these industries contracted by 12.3%. But the decline ranged

Table 13.1 Percentage regional distribution of total employment in New IT producing industries, GB, 1881–1981.

	1881	1891	1901	1911	1921	1931	1951	1961	1971	1981
South-East	—	—	—	74[a]	48	71	56	60	52	54
North	—	—	—	—	3	0.8	4	3	5	4
Yorkshire and Humberside	—	—	—	2	3	2	3	2	2	2
East Midlands	—	—	—	0.4	3	4	5	5	5	5
East Anglia	—	—	—	0.1	0.6	2	3	2	3	3
South-West	—	—	—	0.3	0.7	1	4	4	6	7
West Midlands	—	—	—	5	19	9	10	6	7	7
North-West	—	—	—	17	14	9	11	9	10	7
Wales	—	—	—	0.1	5	0.5	2	2	3	3
Scotland	—	—	—	—	4	1	2	7	11	8
Great Britain[b]				1069	83 875	78 819	275 036	484 140	577 610	484 670

Source: Census of Population (various) NOMIS.

Notes:

[a] Percentages may not add up to 100 due to rounding error.

[b] Total employment.

1981 total from Census of Population; regional breakdown from 1981 Census of Employment.

Table 13.2 Employment change in New IT industries, GB, 1971–81.

Region	1971 employment	1981 employment	1971–81 change	% change 1971–81
Wales	16 419	16 075	−344	−2.095
West Midlands	37 409	36 276	−1133	−3.029
Yorkshire and Humberside	11 912	10 680	−1232	−10.343
South-West	36 582	35 301	−1281	−3.502
East Anglia	16 440	13 252	−3188	−19.392
East Midlands	26 020	22 277	−3743	−14.385
North	25 807	19 046	−6761	−26.198
Scotland	47 042	40 018	−7024	−14.931
North-West	52 700	36 101	−16 599	−31.497
South-East	291 166	263 606	−27 560	−9.465

Source: NOMIS.

from as little as 2.1% in Wales or 3.0% in the West Midlands to a staggering 31.5% in the North-West. The contraction in the South East, at 9.5% or a little less than the national average, still represented a loss of over 27 000 jobs: some 40% of the national total. Still, in 1981, the South-East had 53.5% of the country's New IT jobs, a figure 1.58 times its share of all employment. No other region, in fact, recorded a Location Quotient for New IT above unity (Table 13.3). By individual New IT industries (Table 13.4) the South-East's share was lowest, at 30%, in MLH 363, telegraphic and telephone apparatus and equipment; in other sectors it ranged from 49% (354, scientific and industrial instruments) to 72% (MLH 365, broadcast receiving and sound reproduction equipment).

The conclusion is clear: New IT industries have remained locked into the core metropolitan region of the country: the London region did not suffer the fate of Greater Berlin (Fig. 13.1). But within it, as the following historical account will show, since World War II they have decentralized from London itself to the outer rings of the region (Fig. 13.2).

In Chapter 8 we saw that in the UK before World War II, the electrical engineering industry – precursor of today's electronics – had strongly concentrated itself in London; but that within London it had increasingly decentralized from inner-city core to suburban periphery, especially within a north-west sector already identified by Smith in his survey of 1933. After the war, the 1950s rearmament drive gave a huge boost to the development of military electronics. The firms that profited were those which had already begun to move into the field during World War II. As they grew, they were forced to seek larger sites. But these were no longer

Table 13.3 GB regional distribution of employees in employment in New IT industries (1981) ranked according to location quotients.

Regions	Total employees			Male employees			Female employees		
	Order	LQ 81	No.	Order	LQ 81	No.	Order	LQ 81	No.
South-East	1	1.58	263 606	1	1.66	180 350	1	1.43	83 256
South-West	2	0.98	35 301	2	1.03	24 277	7	0.83	11 024
Scotland	3	0.87	40 018	3	0.84	24 881	6	0.93	15 137
East Anglia	4	0.85	13 252	4	0.77	8104	4	0.98	5148
West Midlands	5	0.77	36 276	6	0.68	21 365	5	0.95	14 911
Wales	6	0.74	16 075	9	0.48	6944	2	1.25	9131
North	7	0.73	19 046	8	0.58	10 133	3	1.02	8913
East Midlands	8	0.66	22 277	5	0.68	15 424	9	0.60	6853
North-West	9	0.64	36 101	7	0.65	23 813	8	0.61	12 288
Yorkshire and Humberside	10	0.25	10 680	10	0.26	7518	10	0.22	3162
total			492 632			322 809			169 823

Source: NOMIS.

Table 13.4 GB regional distribution of total employees in employment in New IT industries (1981) by MLH ranked by total employment size (% by region).

Region	MLH 338	MLH 351	MLH 354	MLH 363	MLH 364	MLH 365	MLH 366	MLH 367	Total
South-East	65	51	49	30	47	72	61	64	54
Scotland	9	1	8	5	8	1	13	10	8
West Midlands	5	3	6	25	7	1	7	3	7
North-West	6	1	6	12	8	2	11	6	7
South-West	5	39	12	1	7	3	4	7	7
East Midlands	5	—	4	9	6	2	1	4	5
North	1	2	3	10	7	3	1	1	4
Wales	—	2	2	6	4	8	2	2	3
East Anglia	1	2	6	1	4	6	1	1	3
Yorkshire and Humberside	3	1	4	1	2	2	1	2	2
Total	100	100	100	100	100	100	100	100	100

Source: NOMIS.

Note: Percentages may not add up to 100 due to rounding error.

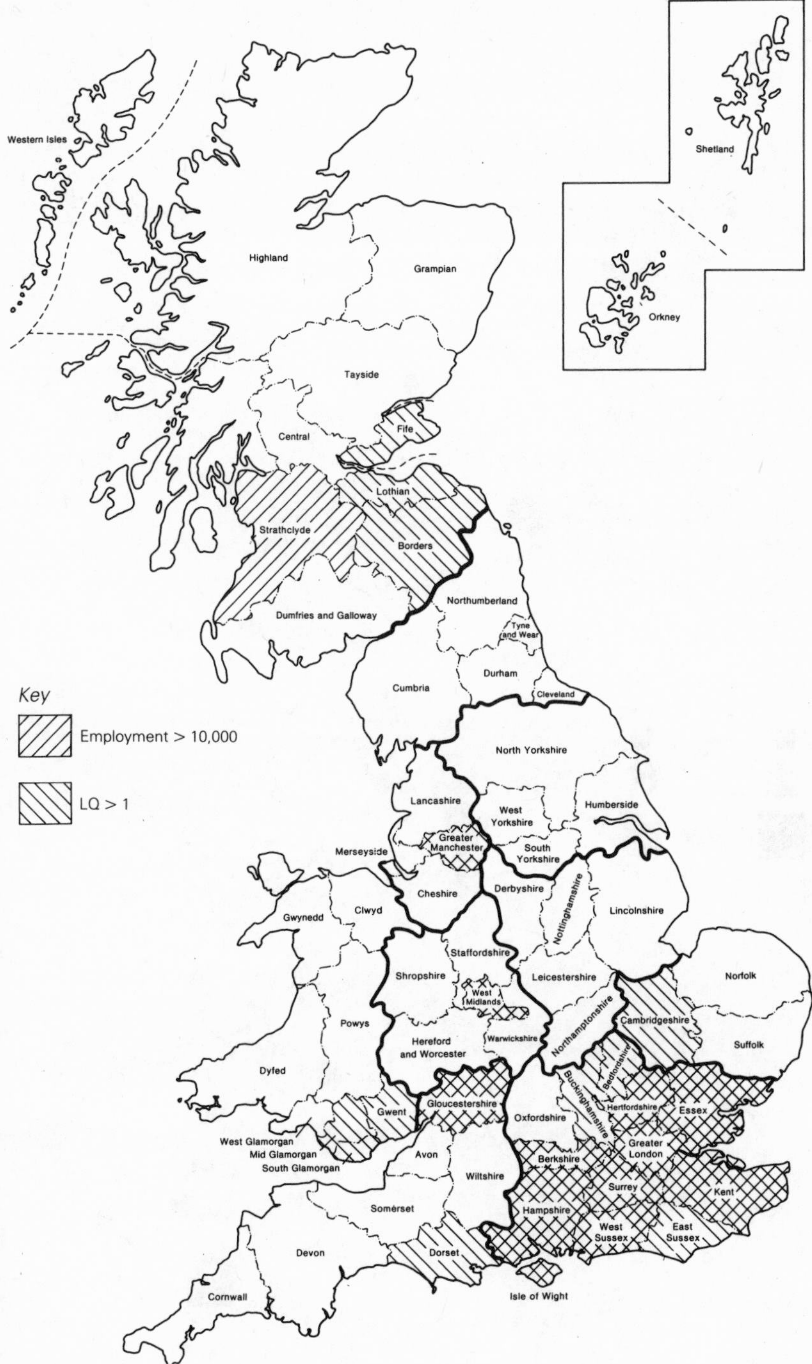

Fig. 13.1 New IT industries, employment location, UK, 1981

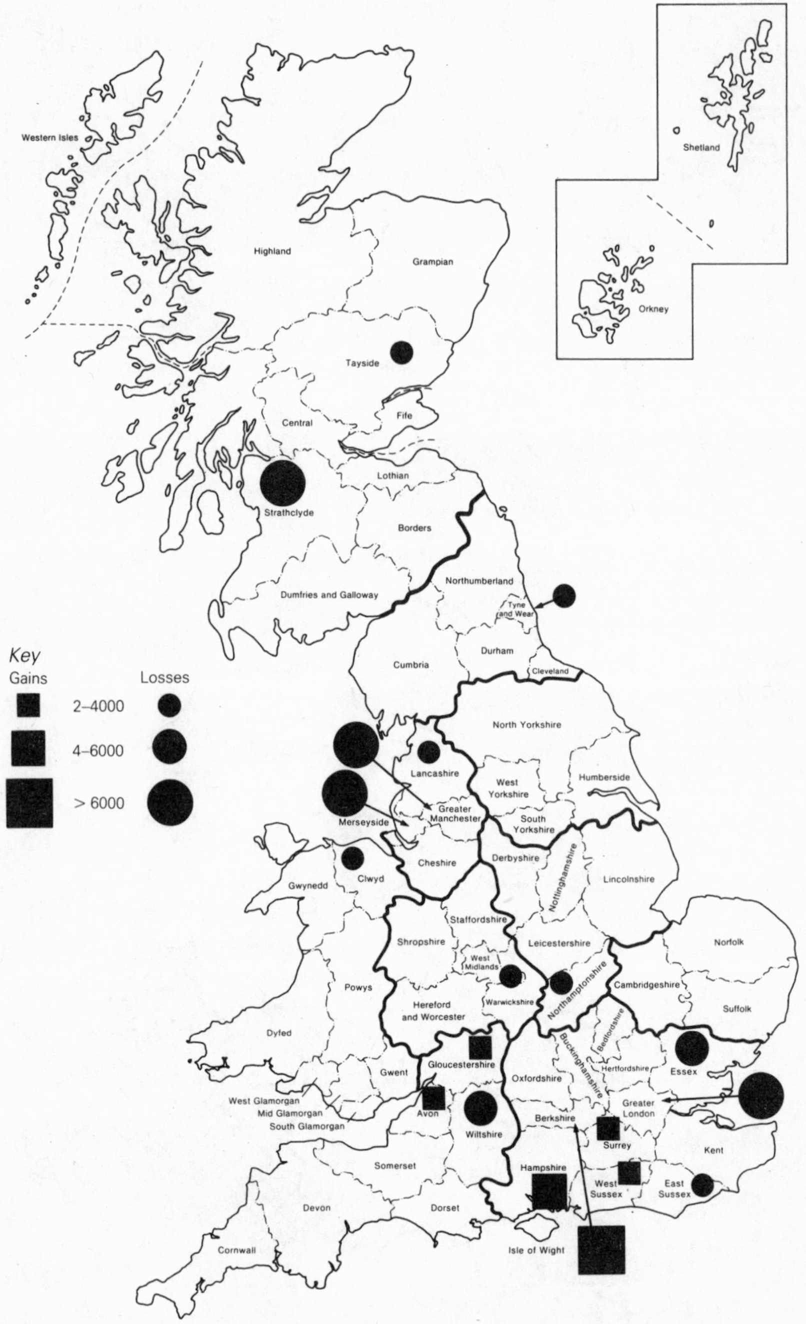

Fig. 13.2 New IT industries, employment location gains and losses, UK, 1971–81

available in London, because green belt restrictions were by then making peripheral factory-building impossible. The firms therefore leapfrogged the green belt, the majority relocating in sites up to 60 miles (100 km) distant, and within that belt mainly north-west of London. Thus, David Keeble concluded in his survey in the late 1960s, the same forces that propelled London's industry in the interwar period continued to drive it; the only difference was that the planning system now took it further afield and then concentrated it into a relatively few nucleated settlements in the same north-west sector (Keeble 1968, Hall *et al.* 1987).

This, however, is not a complete explanation; for many of the interwar firms were located due north of London, whereas the postwar expansion had a distinctly western bias. The explanation is provided by the Government (more strictly, Defence) Research Establishments (GREs), which had already clustered west of London before World War II but greatly expanded during and after it, particularly during the 1950s. These establishments seem to have followed a pattern of incremental growth; they started life as offshoots from existing military establishments in traditional locations, and continued to grow either there or close by. Thus the naval establishments grew in and around Portsmouth, home of the Royal Navy since Tudor times, or in Weymouth; the Royal Aircraft Establishment, biggest and most industrially significant of all, was originally established as a military balloon school and factory in nearby Aldershot in 1892 because this was the home of the Royal Engineers. Others, such as the Atomic Energy Research Establishment at Harwell in Oxfordshire, came into being in World War II after a fairly hasty search constrained by the need to locate west of London for security reasons (Hall *et al.* 1987).

Underneath this apparently random and incremental history, though, there is a deeper logic: London was the centre of the military and scientific establishments; the main defence concentrations, such as Aldershot and Portsmouth, were located to the west for strategic reasons; scientific research from the start of the 20th century was overwhelmingly concentrated in London, Oxford and Cambridge. Even the availability of surplus royal residences, the product of the hunting pastimes of Tudor and Stuart monarchs in the favoured zone west and south-west of the capital, played its part at critical points in history – as in the location of the National Physical Laboratory in Bushy Park in 1900, or of the Radio Research Station in Ditton Park in the 1920s. All pointed to sites west or south-west of London: the future 'Western Corridor'.

The real significance of this location came with the growth of the UK's military–industrial complex in the 1950s. At this point, extremely close relationships began to develop between the GREs and certain industrial firms who did contract work for them. This was particularly true of the leading electronics firms (English Electric, Edison Swan, General Electric, Mullard). Because of the need for close day-to-day ties during the development and subsequent execution of the contract, these firms tended to cluster around the GREs and the Ministry of Supply (later, Ministry of Defence) headquarters in London. The result was that by the late 1970s over 40% of all defence procurement expenditure went to the South-East region – a level of spending that quite dwarfed regional aid, ironically designed to promote peripheral regions at the South East's expense. Of the 57 UK-based contractors who received £5 million or more for defence equipment in 1981–2, no fewer than 43 had their head offices in the South East, Firms which contracted for defence work, interviewed on this point, were quite plain about the need to remain close to the source: managers in one of them said that if they moved 100 miles away from London visits from the military might fall by half. Exaggerated or not, clearly it was believed. Particularly in the early stages of developing a contract, face-to-face contact appears to be vital. And, once a prototype is tested and built, the tradition has been to give the production contract to the same firm.

The British experience is interesting because it strongly suggests that in the development of New IT agglomeration economies of a quite traditional type may prove crucial. The only new feature is that instead of existing at the scale of an inner-city industrial quarter they now exist at the scale of the metropolitan region. We have already noticed similar agglomeration economies in the case of 19th-century Berlin; we shall shortly see them recurring in the USA's Silicon Valley and Highway 128. But the UK's Western Corridor shares more in common with Berlin than with Silicon Valley: it still depends critically on access to governmental agencies in the metropolitan region.

The irony of the British case is that, becoming thus locked in to military contracting, many big suppliers and even larger numbers of smaller subcontractors had too little incentive to do anything else. The story of Highway 128 and Silicon Valley during the 1960s and 1970s – during which many military contractors, or their offshoots, began to develop large-scale civil market activities – finds no obvious parallel in the UK. In particular, though the micro revolution of the late 1970s and early 1980s did see the emergence of indigenous British manufacturers such as Sinclair and Acorn, the resulting development was extremely puny in comparison with the explosion of new entrepeneurship in California and elsewhere.

It is for this reason that, as already seen in Chapter 12, New IT in the UK, far from contributing to the regeneration of the economy, has, almost incredibly, contributed its own share to national manufacturing employment decline.

The American case: the Boston–New York Corridor

The American picture, as already observed in Chapter 11, provides a startling contrast. There, in the brief period from 1977 to 1981, 395 000 additional jobs in New IT industries were created: over 58% of the total growth in manufacturing jobs during this period (Table 13.5). But New IT was strongly concentrated in a relatively few key industrial states. Of a total of 1.9 million New IT jobs in 1977, only four states recorded individual concentrations of more than 100 000 (Table 13.6). All these four had Location Quotients above unity, showing that New IT was more strongly concentrated

Table 13.5 USA; New IT jobs, 1977 and 1981.

	1977	1981	1977–81 Change %	
employment, New IT jobs (000s)	1486.4	1881.5	+395.1	(+26.6)
employment, all manufacturing (000s)	19 590.1	20 264.0	+673.9	(+3.4)
New IT as percentage all manufacturing	7.6	9.3	58.6	

Source: US Census of Manufacturing, 1977 and 1981.

Table 13.6 USA: New IT employment, by leading states, *c*.1977.

	New IT employment (000s)	All manufacturing employment (%)	LQ
1 California	298.4	16.9	2.16
2 New York	155.4	10.2	1.30
3 Illinois	121.7	9.4	1.21
4 Massachusetts	588.8	20.1	2.59
5 Pennsylvania	86.7	6.4	0.82
6 Texas	74.1	8.3	1.07
7 New Jersey	73.2	9.5	1.21

Source: US Census of Manufacturing, 1977.

there, relatively speaking, than in the American economy as a whole (Markusen *et al.* 1986).

Particularly notable is the fact that, contrary to popular myth, the New IT industries are not disproportionately concentrated in Sunbelt states. Three of the top four states in the table, in fact, are old-established manufacturing states from the Frostbelt, America's traditional manufacturing heartland. Lower down in the order, too, this persists: Pennsylvania is ahead of Texas, which in turn is only just ahead of New Jersey. Thus only two out of the top seven New IT states are in the Sunbelt. The Boston–New York axis is still a prominent feature of the American New IT landscape; and within it, Massachusetts has the highest relative concentration of New IT employment of any of the top states.

This of course reflects the patterns already established by the third Kondratieff. As already seen in Chapter 8, the centre of American New IT was firmly established during the 1870s in the Boston–New York axis, and this concentration – lightly modified by outward deconcentration – persisted down to the 1940s. Bell Laboratories remained in New York City until it moved to Murray Hill in suburban New Jersey, in 1941, and here the transistor was invented at the end of 1947; RCA's research work was also in New York City, with an outlier at Princeton. IBM's headquarters in Endicott, New York State (about 140 miles north-west of New York City) was the location of one of its three original constituents, the International Time Recording Company, which had been founded in 1900 in New Jersey and had found itself here after several mergers and acquisitions (Sobel 1983). In Greater Boston, MIT research played a key role in the development of radar; work on the early development of the computer was shared among Eckert and Mauchly at the University of Pennsylvania, von Neumann at Princeton, Aiken at Harvard and Forrester at MIT. Eckert and Mauchly had already left university life to set up Univac, the first computer company, in Philadelphia. In 1950 all the signs were that the next wave of scientific–technological advance would occur somewhere within this corridor, probably at a number of points along it. The sole significant exceptions were IBM at Endicott and GE at Schenectady, both essentially local out-movements from the corridor to smaller towns close by (Brock 1975, Noble 1977, Heims 1980, Wildes & Lindgren 1985).

Nor were the indications very different in the 1950s. Bell Laboratories undertook major development work for the US Department of Defense on the transistor, at a special facility at Whippany, New Jersey, not far from Murray Hill (Fagen 1978, Smits 1985). IBM began to take a commanding lead in commercial producer production. Around MIT, the first major electronics complex began to develop from the mid-1950s, perhaps because of MIT's willingness to take military contracts; the area's older firms, like Raytheon, which had already become heavily involved in this work during World War II, expanded.

Three New IT agglomerations: Route 128, Silicon Valley and Orange County

Then new start-up firms developed out of MIT research, or, much more rarely, research at Harvard. Perhaps the oldest, Wang, was started by a Harvard graduate in 1951; Digital, started by an MIT researcher, opened up in 1957; and thenceforward a regular process of breakaway and swarming occurred, with a total of 54 firms founded between 1955 and 1981, 45 of them started after 1965. The whole process was underpinned by the region's huge supply of qualified workers, the availability of venture capital, and the development of agglomeration economies (Fishman 1981, Dorfman 1982, 1983, Wildes & Lindgren 1985). The Greater Boston area, colloquially Highway 128 (the ring highway, along which so many of these new firms located) was thus the first great electronics complex in the USA. One point about it needs underlining: though MIT research and a supply of MIT graduates were crucial for its development, MIT played relatively little conscious role in bringing the phenomenon about, and Harvard played none (Dorfman 1983). Its success lies almost wholly in the ability of a few firms, principally Digital, to identify a new product, the minicomputer, from the late 1960s onward. Although some older firms (Raytheon, Transistron) were involved in early transistor manufacture, they faded away and the area had almost nothing to do with the subsequent development of the semiconductor (Dorfman 1982, 1983).

Thus, despite the comparisons that are constantly made, Highway 128 was a very different phenomenon from Silicon Valley. Highway 128, the original orbital highway around the city of Boston, started in the 1930s but completed only in 1951, links no less than 20 different towns in the Greater Boston area, many of them with old-established manufacturing and service functions (Loria 1984, Saxenian 1985b). The firms established here, as shown in a 1957 survey conducted by MIT and the State of Massachusetts, had mostly moved out from sites close to the city centre: 59% had been located within a 2¼ mile radius of the centre of Boston, no less than 79% within a 4½ mile ring (Loria 1984). Silicon Valley was established in an agricultural area, with only one established city, then medium-sized (San Jose) and a specialized campus town (Palo Alto). The Highway 128 complex grew on the basis of the assembly of a New IT product, the minicomputer; Silicon Valley grew on the basis of production of a component, thence developing downstream into computer manufacture. Though both grew around universities with strong research in electronics, the process was more or less accidental in Massachusetts, closely guided in California.

Nevertheless, the development histories are similar. Highway 128's New IT industries were a spin-off from research at the Massachusetts Institute of Technology (and, to a lesser extent, Harvard University), located in the heart of the Greater Boston area at Cambridge, Massachusetts; a 1960 survey from Edward B. Roberts of MIT showed that more than 175 new start-up firms there had been founded by former full-time employees of MIT or associated institutions (Loria 1984). Linkages continued to be crucial: an investigation at the end of the 1970s by the Massachusetts High Technology Council suggested that 80% of employees had a qualification from one of the 10 Massachusetts universities that now educate students in this area (ibid.) Silicon Valley, similarly, grew out of research on the Stanford campus at Palo Alto.

Frederick Terman, Professor of Electrical Engineering at Stanford, should have made his academic career at MIT, where he had received his PhD, but had been forced by illness to stay in his native Palo Alto. He had already begun before World War II to encourage his graduates to found firms close to campus so as to draw on its research output; two of the first, William Hewlett and David Packard, started in a garage in Palo Alto while they were still students, helped by a loan from Terman, in 1938; in 1942 the Varian brothers pioneered the klystron tube (the basis of modern radar and microwave communications) at the university, then founded Varian Associates nearby. During the war Stanford became a major recipient of federal defence research for electronic components and equipment (Rogers & Larsen 1984, Saxenian 1985c).

Returning to the university from a major military project at Harvard, Terman sought to develop a new university–industry relationship, self-consciously seeking to develop a 'community of technical scholars' in Palo Alto (Saxenian 1985c). Gifts from corporations poured into Stanford; the university developed one of the two best electrical engineering programmes in the country, soon rivalling MIT's and – by the early 1960s – producing more PhDs (ibid.) Stanford Research Institute (SRI) was founded in 1946 with a remit to conduct research to help stimulate West Coast business. In 1951 it was joined by the Stanford Industrial Park, one of the first science parks in the world, called by Terman 'our secret weapon'; it had 7 companies in 1955, 70 by 1970, 90 by 1981 (Rogers & Larsen 1984, Saxenian 1985c). The Korean war, and the Cold War of the early 1950s, boosted defence expenditure on electronics research, in particular through work on the ballistic missile system; Terman attracted a significant share of the Pentagon's research and development programme to Stanford. Lockheed's move to nearby Moffet Field, in 1956, provided a further boost (Rogers & Larsen 1984). The share of military prime contracts going to the region rose steadily from 12.3% in World War II, to 17.9% at the time of the Korean war and to 27.5% by 1960 (Saxenian 1985c).

As a result, major national firms (Sylvania, Fairchild, General Electric, Westinghouse, Itel, Kaiser, Philco-Ford) began to establish operations here. A critical event was the return of William Shockley, one of the three inventors (at Bell Laboratories, New Jersey, in December 1947) of the transistor. He founded Shockley Semiconductors in 1955; two years later, eight of his best scientists broke away to start their own firm, backed by Sherman Fairchild. In 1960 one of these eight, Jean Hoerni, invented the planar process for semiconductor manufacture, which cut costs and brought a mass-production revolution to the industry. By 1965 Fairchild Semiconductors had spawned 10 other semiconductor firms; by 1979, 50, in a process that has produced a unique genealogical tree (Rogers & Larsen 1984).

Thus Silicon Valley grew through a conjunction of two circumstances: a large output of scientific manpower from the region's educational institutions and research institutes; and demand for semiconductors generated by the defence programme. Particularly important, here, was demand from aircraft and aerospace prime contractors and subcontractors, who were the main customers for the young semiconductor firms; Massachusetts lacked this concentration of aerospace contractors (Saxenian 1985c).

Thus the typical Silicon Valley firm was always small; for every one that succeeded in growing large, there would be scores of new formations. One guide to Silicon Valley lists no less than 2736 manufacturing firms alone; over two-thirds have from one to ten employees, 85% have fewer than 50 (Rogers & Larsen 1984). Such small firms depended critically on access to venture capital. The San Francisco area, as one of the leading capital markets of the USA outside New York, offered ready access to funds. There is a strong suggestion, too, that in this free-wheeling innovative climate lenders were more willing to back risky ventures than in older-established regions.

Small infant firms also very much depended on economies of agglomeration for obtaining highly qualified labour, for communicating with their prime contractor and subcontractor customers, for obtaining specialized supplies and services, and simply for access to informal channels of general trade knowledge. Later, during the 1960s, as firms increasingly diversified into civil markets, these agglomeration economies continued to operate and even to gain in relative significance (Rogers & Larsen 1984, Saxenian 1985c). The reason was the critical importance of expertise, which was transitory and very valuable; firms located here could acquire it informally, at the Homebrew Club or at the Waggonwheel Bar, described as the 'fountainhead' of the industry (Braun & Macdonald 1982, Freiberger & Swaine 1984). One measure of this was the continued high rate of new firm spin-off, with 13 such foundations in one year (1968) alone (Saxenian 1985c). Thus the area grew as the high-technology version of a traditional industrial quarter, such as was found in many older cities in the 19th century (Hall 1985). The interesting question is why such an innovative industrial quarter, which once naturally colonized inner-city locations, should now desert it.

A particularly apt illustration of these continuing agglomeration economies is the saga of the birth of the personal computer, already told in Chapter 9 (Freiberger & Swaine 1984). Intel, a Fairchild spin-off located in Silicon Valley, developed the microprocessor – the equivalent of 1950's roomful of circuitry – in 1971. The first true microcomputer, the Altair of 1971, was invented not in Silicon Valley but by MITS, a small firm in Albuquerque, New Mexico. It was essentially a machine for dedicated hobbyists, not for the mass market; its appearance led to the development of the Homebrew Computer Club, 'an extraordinary gathering of engineering expertise and revolutionary spirit from which would spring dozens of computer companies' (Freiberger & Swaine 1984, 103). It first met in a garage in Menlo Park, at the northern end of Silicon Valley, in March 1975, but soon expanded into a large Stanford Linear Accelerator Center lecture hall (Freiberger & Swaine 1984). As its members competed to build the earliest prototype microcomputers, others developed software: CP/M, the first operating system for microcomputers, was developed by Gary Kildall, an instructor at the US. Naval Postgraduate School in Monterey, 50 miles south of Silicon Valley, in 1973; together with John Torode, a friend from the University of Washington, he developed the first disk drive. Finally, Stephen Wozniak (a regular Homebrew Club attender) and

Steven Jobs from Cupertino, in the heart of Silicon Valley, teamed together to produce the Apple, in 1976. They sold it through the local Byte Shop – one of the earliest computer retail shops, which had opened in Silicon Valley's Mountain View the year before (ibid.). The mass-market personal computer was born (Freiberger & Swaine 1984).

The Route 128 and Silicon Valley stories suggest strongly that New IT grows out of organized research in top universities or research institutes. But the third great new concentration of American New IT in the fourth Kondratieff tells a different story. After 1950 Orange County in Southern California expanded from virtually nothing to become one of the major manufacturing centres of the USA, with over 225 000 workers in 1981 – 56% of them in key high-technology sectors like machinery, electrical and electronic productions, transportation (almost exclusively aircraft) and instruments (Scott 1986). The process started with branch plants overspilled from neighbouring Los Angeles during the 1950s, a period of rapid growth occasioned by the defence contracting build-up of those years. Then, in the 1960s, it developed into an industrial complex of small firms, characterized by dependence on a common labour pool and common infrastructural services.

Allen Scott concludes that Orange County 'is a major growth center in the classical sense of the term . . . constituted by a core of dynamic propulsive (high technology) industries around which a penumbra of dependent input suppliers has grown up [and underpinned by] a proliferation of a contingent labor force and associated urbanization phenomena' (ibid., 21). Characteristic of such a complex is vertical distintegration of labour processes, which developed as technologies and procedures became standardized during the late 1960s and 1970s: 'The more the complex has expanded, the more it has produced, by its own internal momentum, a wider pool of agglomeration economies, and thus it has expanded wider still' (ibid., 41). Exactly the same conclusions could be drawn for Silicon Valley or for Highway 128; in all three, the rapid evolution of the technology brought a proliferation of new firms, vertical disintegration of productive processes, and consequent economies of agglomeration. Ironically, these were exactly the same features that characterized earlier, lower-technology industrial complexes like clothing in New York or London, or the small metal trades of the English West Midlands (Wise 1949, Hall 1962, Martin 1966). The processes, evidently, are the same.

The difference between the three New IT complexes is this: in Orange County, where the Irvine campus of the University of California was established only in 1965, spin-off from university research could hardly be adduced as the cause. This suggests that the presence of a university, for all its apparent importance in the Silicon Valley and 128 cases, may not always be the catalyst. Indeed, the study by Markusen *et al.* (1986) concludes that when high-tech employment and employment growth in American Standard Metropolitan Statistical Areas are regressed against a number of possible explanatory factors, federal research and development does not prove to have any noticeable statistical association: what does is federal defence expenditure. Apart from this, the factors that prove significant are the ones often assumed to be so in the popular literature. They include amenity variables like climate, a good range of educational options, and a range of affordable housing; access variables like highway networks and good airports; and agglomeration economies like the presence of major company headquarters and a wide range of business services (ibid.). Their list of 100 high-technology industries was much longer than ours of 21 New IT industries, but in 1981 the latter accounted for nearly one-third of total employment in the longer list; so, presumably, the explanatory variables may well be similar.

Conclusion: the geography of innovation

The conclusion that emerges from this comparison is that the geography of innovation shows considerable inertia: once it has taken root, innnovation proves a resilient plant that may locally migrate, but proves difficult to destroy. Berlin, where such a tradition was destroyed through war, may be the limiting case. London, the innovative capital of a nation failing to innovate, exported its traditions to the neighbouring rural counties, where they continued to represent perhaps the only genuine centre of New IT growth in the entire economy. The Boston–New York axis, likewise, continued to show vigorous capacity to develop new technologies, though that capacity came to be located mainly outside the great cities. Silicon Valley was a case of a genuinely new innovative tradition, but even it developed in a major metropolitan area containing two of the USA's major universities, one of which proved directly instrumental in its birth and nurture.

Of these case studies, Greater Boston is perhaps the most significant for students of regional and urban policy. It is indisputably the USA's oldest major city, and its first seat of higher learning. It draws on a 350-year tradition of education and scholarship, and a 140-year history of work in advanced technology.

It was the main centre of technical innovation in New IT in the 1860s and 1870s, even though inventors like Edison were drawn to the New York marketplace. It continued to maintain its lead in basic scientific research in the key fields of electronics throughout the critical years of the third Kondratieff depression. Thus it was poised to take the lead at the start of the fourth. Boston shows that an old, indeed very old, industrial region can remain innovative. But it depends on the maintenance of the right kind of innovative milieu, which demands a delicate balance between fundamental research and commercial exploitation. Silicon Valley created that milieu and that balance; Greater Boston retained and further developed it.

PART FIVE

The information age

Convergent IT in the fifth Kondratieff, 2004–

14
Explanations and speculations

We have reviewed a century and a half of economic history in order to ask two central and related questions: How, and why, does the rate of new technology based industrial innovation vary from decade to decade, and how does its locus shift from country to country and (a subsidiary question) from region to region? We have used the so-called New IT industries as representative examples of innovative products, which have constantly adapted so as to remain at the forefront of technological advance, thus becoming increasingly important to advanced economies. And, because in the last half-century they have come to depend so fundamentally on electronics (more recently, microelectronics), we have separately traced the evolution of electronic out of electrical engineering.

As an analytical device, we have adopted a Schumpeterian framework based on Kondratieff long waves of industrial innovation (Fig. 14.1). In the course of three such long waves we have seen extraordinary shifts in the world pattern of New IT production: the early German and American challenge to the UK at the very birth of New IT during the second Kondratieff; the parallel rise of Germany and of the USA at the opening of the third, with the failure of the UK to build on its early lead; the dominance of the USA and the challenge of Japan, coupled with the virtual extinction of the UK as a New IT power by the middle years of the fourth.

The rôle of basic innovation

There are as many possible explanations of these changes as there have been analysts, which means a great many indeed. As we shall now seek to show, they are not mutually exclusive. These explanations must deal simultaneously with two phenomena: the historical timing of the advent of major new industries, and the geographical location of such developments. A majority of researchers now seem to accept, as a first level of explanation, that the timing is systematically related to bunching of industrial innovations – the conversion of original inventions into commercially marketable products – in certain phases of the Kondratieff cycle; they continue to disagree on the deeper causes of this process. Almost all observers, including ourselves, appear to accept that the geography of change, both for nations and regions, is directly related to their capacity for such original innovation, or, at least, to their ability to borrow it through technological transfer.

Beyond that, agreement tends to stop. In Chapter 2 we reviewed the debate and outlined our tentative approach, which follows recent interpretations by Freeman, Perez and others: what is significant for the origin of each long wave is not swarms of individual innovations *à la Mensch*, but rather chains of related innovations ('technology systems') which stem from one or two major technological innovations; in this process, organizational and institutional innovations may prove important; in them, state action as well as sociocultural factors may play a crucial rôle; because the adoption of such systems is a very complex process, involving major adaptations in the wider society – in its ability to generate the entrepreneurial figures, in the training and aptitudes of its workforce, in its capacity to develop patterns of consumption, in the ability of the state to provide the necessary infrastructural and regulatory framework – it inevitably takes time.

This, we believe, was Schumpeter's own interpretation when he wrote of the advent of 'New Men' at the start of each Kondratieff (Schumpeter 1939); it was also the approach of Schumpeter's disciple François Perroux (1961) in his much misunderstood concept of the growth pole:

> . . . la croissance n'apparait pas partout à la fois; elle se manifeste en des points ou des poles de croissance avec des intensités variables; elle se répond par divers canaux et avec des effets terminaux variables pour l'ensemble de l'économie.

The 'poles' are not, as commonly assumed, geographical clusters but are industries embodying key innovations; the 'channels' may be partly silted up by the accretions of the social and cultural and political past; so the process everywhere takes years or decades, and in some places may be fatally impeded.

It goes almost without saying, further, that it is difficult to measure innovations. The lists do not agree as to contents or to dates or to national origins, or even as to what constitutes an innovation. If our argument about technology chains and social innovation has merit, the rate of diffusion is at least as significant as the date of the initial innovation.

Nevertheless, our first conclusion is that by using judgement in selecting what appear to be key innovations, these do show a strong tendency to occur in Kondratieff downswings (more precisely, in Kondratieff recession phases) as theory suggests they should. Table 14.1 is based on van Duijn's list of significant New IT innovations. Although all have had economic significance, we would judge only a few among them to have been truly key technologies of the kind that engender a whole innovation chain and thus bring about a major new technical–industrial system. Table 14.2 is based on Dummer's listing of such innovation chains in electronics. They extend over long periods – sometimes, more than one Kondratieff. From a comparison of the two tables, we can produce a schematic

Table 14.1 Major New IT innovations: dates and countries of origin.

electric telegraph	1939	UK
atlantic telegraph cable	1866	USA
typewriter	1870	Denmark
telephone	1877	USA
halftone process	1880	USA
punched card	1884	USA
cylindrical record player	1888	USA
mechanical record player	1889	Germany
vacuum tube	1913	USA
AM radio	1920	USA
dynamic loudspeaker	1924	USA
electric record player	1925	USA
radar	1935	Various
magnetic tape recording	1935	Germany
FM radio	1936	Germany
television	1936	UK
phototype	1946	USA
long-playing record	1948	USA
xerography	1950	USA
transistor	1951	USA
electronic computer	1951	USA
colour television	1953	USA
silicon transistor	1954	USA
integrated circuit	1961	USA
communication satellite	1962	USA, USSR
video cassette recorder	1970	Netherlands
microprocessor	1971	USA

Source: van Duijn (1983).

Table 14.2 Electronics innovation chains.

Original invention	Sample successor innovations	Number of successor innovations	Dates of chain
phonograph	LP record stereo record compact disc	12	1877–1982
telephone	PBX MODEM STD digital systems	13	1876–1975
wireless telegraphy	broadcasting FM packet switching	11	1896–1976
Two-electrode valve	klystron MASER	14	1909–74
Electronic TV	TV broadcasting	4	1919–36
digital computer	IBM 701 minicomputer microcomputer Cray computer	21	1939-79

Source: Dummer (1983).

picture of the timing of key innovations in New IT in relation to the conventional Kondratieff chronology and Mensch's 'radical years' of innovation surges (Fig. 14.1).

Five technology systems, and their diffusion

We would argue that at least five, possibly six, critical technological chains can be distinguished for New IT and related electrical technologies. The first came with the original New IT innovation, the telegraph, in 1837–9, which represents the end of Mensch's 1825 innovation peak corresponding to the second Kondratieff downswing; aided by the development of a subsidiary technology, the undersea cable, it duly generated a second Kondratieff industry, the significance of which was mainly as a subsidiary to the key second Kondratieff industries of railways and steamships. Because of the novel technological challenges it posed and huge scale of the

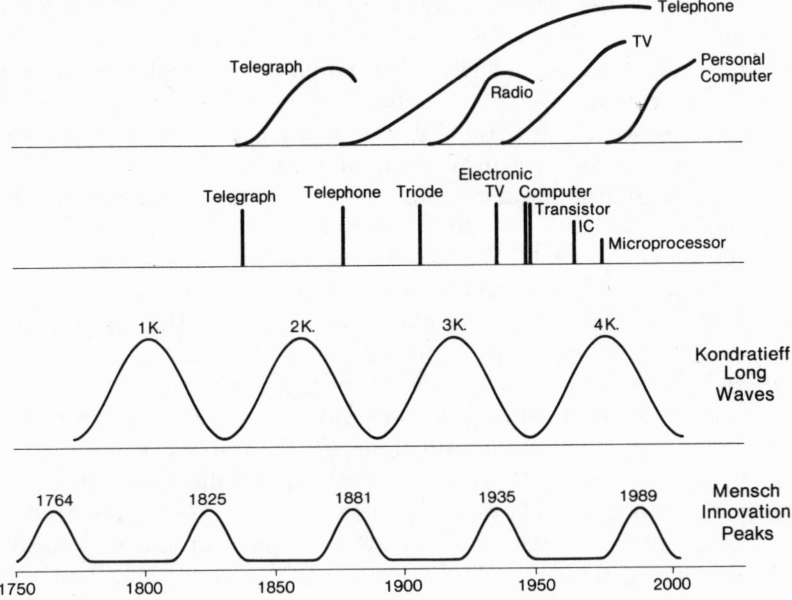

Fig. 14.1 Schematic diagram of radical New IT innovations and the Kondratieff long waves

investment involved, however, it took much of that Kondratieff for the technology to diffuse.

The next link in this chain, the telephone, came in 1876, at the start of the 1881 innovation wave (corresponding to the second Kondratieff downswing); it generated a major third Kondratieff industry. The basic supplementary technologies (the exchange, the switching system) were in place just over a decade later, though successful very-long-distance transmission had to await the birth of electronics some two further decades after that. And, more importantly, the necessary social, cultural and political innovations were so momentous that the true impact of the telephone diffused only slowly; its rapid acceptance and spread in the USA contrast with its difficulties at the hands of postal monopolies in Europe. So, like the automobile, its universal diffusion occupied at least two entire Kondratieffs. Thus far, however, the initial innovation chronology neatly follows the Mensch hypothesis, with a double Kondratieff chain in the case of the telephone.

The second major chain, consisting of a cluster of mechanical New IT technologies of the period 1870-90 (the typewriter, phonograph, duplicating machine and accounting machine) also neatly spans the Mensch 1881 innovation peak. This chain, furthermore, is a relatively short one. Perhaps because the technologies required no new and complex infrastructure to link them – all that was needed was long since in place, in the form of the postal system – the diffusion was relatively rapid and painless, albeit again much faster in the USA than in Europe. Some parts of this technological system were however transformed by the application of electricity – to the record player in the 1920s, to the typewriter in the 1930s.

The third chain, representing the major electrical innovations of 1870–90, also coincides particularly well with the Mensch scheme. In technological terms it represents a short chain, started by a swarm of innovations that almost simultaneously provide a way of generating and transmitting power, and a number of uses for it; it was followed about a decade later by an improved alternative system for the former. But its diffusion entailed the development of complex local, regional and then national networks of transmission and distribution, which required the development and imposition of uniform technical standards, a process that took time, and – at least in the case of the UK – was badly bungled, with incalculable consequences for British competitiveness in the new industries. Thus the economic impact was attenuated, effectively, over an entire Kondratieff; not until the 1950s was electricity effectively brought to every corner of the major industrial countries, and not until even later was the average home comprehensively equipped with electrical appliances.

The fourth of these technology chains, however, represents an important divergence. As shown in Chapter 6, the first true electronic technologies (Marconi's experimental work, the diode of 1904 and the De Forest triode of 1906) fall right outside any Mensch innovation peak, in the first decade of a Kondratieff upswing. This was because they formed an indirect and late link in the electrical-technology chain. (If Edison had been less desperate to concentrate on commercially useful inventions, we can speculate, he would have pondered over his Edison effect, and the electronic age would have been born, in 1882 – just right on the Mensch chronology.) It immediately generated the first electronic industry, radio. Starting as a producer-good industry – like the telegraph, used for shipping and military purposes and massively developed for the latter during World War I – this then, after the organizational innovation of broadcasting in the 1920s, span off a large consumer-good industry, which developed in the third Kondratieff downswing and proved a powerful countercyclical force. Once again, the new technology diffused fastest in the USA where, developed by private enterprise, broadcasting acquired a head start; the Europeans, who preferred the route of the public corporation, were a few years behind, though the British Broadcasting Corporation co-operated with the major electrical producer to pioneer public television broadcasting in 1936.

The fifth major bunch of electronic and electric New IT innovations all fell fairly within the span of Mensch's 1935 innovation peak (though Dummer's dates, which refer to invention, require careful interpretation): they included sound films, electronic TV, FM radio, xerography, high-fi recording, radar, the computer and the transistor, all coming between 1935 and 1952. Viewed more closely, they really form two separate groups: one, based on innovations between 1935 and 1948 in the field of traditional valve-based electronics and electrical technology; the other, starting in 1947 and forming a chain through the 1950s and 1960s, constituting the new microelectronics and related computer technologies. Wartime R&D, we surmise, may have speeded one of these basic innovations, the computer, though it may have delayed the other, the transistor.

At the start of the fourth Kondratieff, around 1950, these two sets of innovations logically produced two separate streams of industrial development: the first in consumer-good industries (TV, high-fi, car radio), largely developed by established firms and forming an important contribution to the start of the Kondratieff upswing; the second, developing more slowly through a mixture of established and new firms, in producer-good industries for commercial and military applications (mainframe computers, ballistic missiles and spacecraft).

The first of these required no special infrastructural provision, since just at this point the electrification of the advanced countries was complete; it did, however depend on socio-institutional innovation in the form of the mass-consumption (Fordist) society, driven by the expansion of consumer credit. The second, in contrast, entailed massive and extremely expensive infrastructural investment in the form of space stations, launching pads, and improved communications technologies. In both branches, but especially in the latter, a long chain of further product and process innovation took place: colour television, cassettes, digital recording and compact discs in the first; integrated circuits, microprocessors, calculators, personal computers and new telecommunications services in the second.

In the latter group of industries the result was a dramatic shift down the learning curve – greatly assisted by military subsidization of the early R&D – to ever lower-cost mass production. Further, the basic component innovations were combined and applied in a range of increasingly sophisticated products such as calculators, personal computers and digital compact discs. These came on stream only in the 1970s and 1980s, just as the Kondratieff upswing was giving way to downswing. Although component production was increasingly dominated by large firms which could afford the increasingly high capital costs, their embodiment in new products came through a mixture of older-established firms and new entrants, with much vertical disintegration of the productive process.

There is a curious parallel here with the history of radio in the 1930s. As we have already seen, the three-electrode De Forest valve of 1906 came in the first decade of the third Kondratieff, and was a logical (though inexplicably delayed) result of basic electrical innovation 30 years before. Its impact delayed by a world war, it resulted in a major consumer-product industry 15–25 years later. The microprocessor of 1971 came in the second decade of the fourth Kondratieff (by Schumpeterian chronology) and was the logical result of a continuous innovation chain that had started with the transistor; it was incorporated into commercial products, notably the personal computer, less than a decade later.

This might suggest that the fourth Kondratieff was wholly an electronics-driven Kondratieff. That is far from true. A large part of that Kondratieff was clearly driven by the extended life cycle of third Kondratieff industries such as automobiles and domestic electrical goods, thus proving Schumpeter's argument that key innovations can pervade more than one Kondratieff. In particular this was due to the diffusion of these technologies from the USA, where virtually all of them had originated, to other parts of the advanced industrial world, above all western Europe and Japan. In absolute terms, for instance, world production of cars in the fourth Kondratieff far exceeded that in the third (Altshuler *et al.* 1984). But we think it correct to say that electronics, in particular microelectronics, were one of the true 'carriers': they represented the distinctively new technologies of that Kondratieff, as cars or electrical goods were in its predecessor. Their quantitative impact in its early decades was necessarily small; it became noticeable only towards the end of its prosperity and into its recession phase, just like cars or radio in the third Kondratieff. But – again as in previous Kondratieff upswings – it was almost immediately multiplied by indirect multiplier effects, particularly through the necessary construction of infrastructure, the development of service delivery industries, and related urban construction in the regions and cities where the new industries were concentrated.

The rôle of infrastructure

Indeed, there are intriguing parallels with the history of innovation in other periods, both in New IT and in other fields. One critical point is that much innovation proves to depend for its exploitation on the creation of an infrastructural network (railways; telegraph and telephone; electricity grids; highways; airports and air traffic control; telecommunications systems). As Jonathan Gershuny has put it, in discussing the traditional Keynesian countercylical multiplier:

> It is the particular performance characteristics of the new infrastructure that combine with new technologies to *enable* the diffusion of the mass consumption goods. Spend money on the National Grid, and we complete a technological paradigm that leads to the vast new markets for domestic electrical equipment; dig a very large hole in the ground, and we have the new electrical and materials technologies still lying economically fallow. Thus, quite distinct from the rôle of infrastructure investment in the operation of the Keynesian multiplier, is its rôle in a sort of super–multiplier, the stimulation, or *enabling*, of specific new products for innovative household electrical products. (Gershuny 1987, 496).

This, almost by definition, has to be achieved comprehensively and in a relatively short time; it requires very large-scale organization and massive capitalization; it also requires technical standardization (railway gauges; standard times; voltages; communication protocols; rules of the road; speed limits; driving tests; motor insurance; international air traffic control; computer reservations systems). We invariably find therefore, that the infrastructural consequences of major innovations involve the creation of very large-scale, vertically integrated enterprises, often the largest ever seen down to their day, sometimes – even in the *laissez-faire* era – requiring state provision, or at least major state concessions (the early railways; the electrical utilities; the telephone companies; the airlines).

The consequent problems of standardization and co-ordination, across nations and between nations, inevitably delay the diffusion of the new technologies, sometimes for decades: in the second Kondratieff, the UK suffered from the battle of the railway gauges; in the third, the UK was similarly plagued with inconsistent electrical systems, while Europe was confusingly divided between right-hand and left-hand rules of the road and by cumbrous restrictions on frontier crossings by road; in the fourth, different national telecommunications standards inhibit the free transmission of data. Eventually, complex systems of state regulation or international protocols are developed to overcome these problems; but the resultant delays may be critically important in postponing the diffusion of the new technologies, and so in inhibiting demand for the resultant products. In particular, areas with large territories under common jurisdiction and with common standards (Germany from 1871 to 1945; the USA throughout its history) will have a real advantage in diffusing new technologies over areas plagued by multiple standards and barriers to communication (Western Europe both before and – despite the Treaty of Rome – after World War II). What matters here is not merely the lack of conventional tariff and other trade barriers, but common technical standards – especially for the exchange of goods and people in the third Kondratieff, of information in the fourth.

Many observers now think that Convergent Information Technology (the carrier of the fifth Kondratieff) will demand a similar huge infrastructural investment. They point to the Japanese investment in the Integrated Network System (INS): a fully integrated broadband network based on high-capacity fibre-optic cable, which by the mid–1990s will be capable of carrying essentially all communications via a single network (Mackintosh 1986, Cohen & Zysman 1987, Forester 1987). Such a system also demands that all equipment be constructed so that it 'talks' to every other: a requirement necessitating an extraordinary degree of co-operation between those building the infrastructure and those making the equipment to use it. It will, Cohen and Zysman argue – and we agree – have as profound an effect on the structure and location of production as railways, motorways and jet aeroplanes did in their eras. The evidence is that Japan is far in advance of any other country, both in building the infrastructure and in developing the structure of ownership and control needed to make it work: a structure embodying competition between Nippon Telephone and Telegraph, the major Japanese equipment manufacturers, and outside competitors, all within a structure of policy guidance from the Japanese Ministry of International Trade and Industry (MITI) and Ministry of Posts and Telecommunications (MPT). No other country, so far, seems to have approached these questions in a similarly coherent way.

Economies of scale in New IT

Almost definitionally, therefore, infrastructure systems demand very large-scale private corporations or state utilities. But some part of the resulting production, too, may be extremely capital intensive and open to massive scale economies, leading similarly to the development of huge vertically integrated companies (Bessemer steel, the electrical cartels, the major automobile companies). In some cases, however, the productive process does not involve such massive capitalization, so that, provided patent rights can be utilized or circumvented, new small firms may enter and the productive process may be vertically disintegrated: thus radios and components in the 1930s, personal computers in the 1980s.

Quite often, significantly, these products prove to be independent of the existence of an infrastructural grid, or, more precisely, independent of the need to create a *new* grid: examples are typewriters and gramophones in the third Kondratieff, colour television or personal computers in the fourth. (Typewriters depended on existing postal systems, personal computers on existing electricity grids.) Even then, it is true, entrepreneurial inexperience may lead to a high rate of mortality among the new firms, resulting in sharp market concentration after a few years (radios and personal computers again). But the distinction between the two kinds of industry does seem to be a very basic one in the history of innovation.

The geography of innovation

Table 14.1 also confirms the changing international geographical pattern of innovation: the steady decline of the UK as an innovative power, the rise of Germany and then its abrupt fall after World War II, the rise of the USA to world technical leadership. There is a British lead in the first Kondratieff, with continuous and remarkable American dominance thereafter: a dominance even more marked for New IT than for innovation in general. During the second and subsequent Kondratieffs, the USA accounted for between 67% and 100% of major New IT innovations that can be ascribed to a single country. The only significant challenge comes from Germany, but in the fourth Kondratieff this almost disappears. The steady decline of the UK, from each Kondratieff to the next, is only too evident; the British competitive failure can be traced directly to the country's failure to innovate (technologically and organizationally), and it goes back a long way.

There is, however, one clear anomaly, which is not unexpected: it is the failure of Japan to appear in the lists. Not only basic inventions but also first commercial applications are made elsewhere, to be adopted and perfected by Japan: a confirmation, apparently, of the popular myth. This, however, would be wrong; it points to a basic difficulty with all the innovation data. Necessarily, they show the first commercial breakthrough, but entrepreneurial history shows that this may not be the same as the ultimate commercially successful application. Edison developed an electric typewriter in the 1870s, but IBM made it commercially successful only in the 1950s. Colour television was perfected in the USA in the 1950s, but the Japanese mastered high-volume, high-quality, low-cost production in the 1970s. Various companies developed low-cost domestic videorecording systems in the 1970s, but the successful company was a late entrant, the Japan Victor Company. Numerous other examples could be given. The point is that for most of the fourth Kondratieff the Japanese have proved highly adept at perfecting a version of a new product that meets market needs, and then developing high-volume, high-quality, low-cost production: a combination of improvement product and process innovation that has proved unbeatable. In these circumstances it is a moot point as to the identity of the real innovator. But in more recent years, Japanese firms have begun to take a lead in initial innovation in some key IT product fields.

Table 14.1 is also important in explaining the relatively weak British performance in both the third and fourth Kondratieffs. For it disproves a common assertion: that the British have been successful in original innovation but have been weak in follow-through. In some cases, true, early scientific advances and technological inventions may have occurred in countries other than those that first made the successful commercial application; there is little reason to believe that the UK shows up better in basic inventions in the IT sector. But, in terms of application, the British economy has shown a decided competitive weakness since the late 19th century.

Innovation and industrial organization

The changing locus of innovation, though, provides only a first level of explanation. The next question has to be why innovative potential should shift in this way. Here, abundant evidence from our study suggests that the nature of innovation has indeed changed, as Schumpeter (1939, 1942) had recognized. Down to the start of the third Kondratieff, primary innovations could still be made by relatively small firms which in effect created new industries, where they achieved a kind of instant leadership. This, furthermore, might happen with minimal state aid or none at all.

Though Siemens owed much to the support of the Prussian telegraphs when he started in Berlin in the 1840s, at the outset of the second Kondratieff, Edison owed nothing to the American government (save, perhaps, for its indirect rôle in opening up the country) when he achieved his extraordinary burst of New IT and related electrical innovation in the downswing of that long wave during the 1870s. Even though the German government played a rôle in the development of the new electrical technologies there in the 1880s, this rôle was generally restricted to regulation coupled with indirect technical assistance, admittedly significant, in the form of support for education and to its mercantilist industrial policies. And, though the German government was actively supporting the early development of radio even before 1914, it was not until the end of World War I, when the federal government helped engineer the birth of RCA as a rival to the British Marconi undertaking, that we can trace any systematic state promotion of New IT innovation in the USA (Maclaurin 1949, Sturmey 1958, Wilson 1964, Freeman *et al.* 1965). Down to the mid-point of the third Kondratieff, the state's rôle in innovation was thus regulatory and facilitative, no more.

What did happen, from the start of the third Kondratieff, was a quantum jump in the scale of high-technology industry. The new electrical and telephone firms of the 1880s and 1890s (General Electric, Westinghouse and Bell in the USA; AEG in Germany) were giant enterprises almost from the start. They resulted from an organizational innovation, the development of close linkages between industrial and finance capital, and they exploited their patents (not infrequently, buying or trading them) to achieve market power; not by accident did the electrical age immediately become the age of the cartel. This was because of the scale of organized R&D necessary to ensure commercial leadership in a complex new high-technology system, coupled with the capital-intensive nature of much of the production and the economies of scale that could be achieved thereby. In the early 20th century these companies helped develop the infant electronic industry. Yet – an important point – in several critical instances they did so slowly and reluctantly; new entrepreneurs, like Marconi in Great Britain, could enter the field at this time and win a significant market share. In the other main stream that eventually fed into New IT, the mechanical and electromechanical office technologies, many firms, some established, some new, at first fought for supremacy. This was equally true of typewriters, duplicating machines, dictation equipment, tabulating and accounting machines. In all, new entrepreneurs could and did make their mark during the 1880s and 1890s, creating new commercial empires. Nowhere was this more evident than in the USA.

Even later, the triumph of the large oligopoly was, however, by no means complete. Successful innovation-based growth depended on more than hardware; it also needed a range of other forms of specialized knowledge and skills in design, marketing and production control as well in the provision of venture capital, which are too often neglected. These might be achieved either by vertical integration of research, planning, production and marketing functions in a single large firm, or by vertical disintegration into many small specialized firms linked to each other in a single industrial agglomeration. The American New IT sector offers contrasting examples of both models: IBM on the one hand, the many smaller firms of Silicon Valley or Orange County on the other. (American anti-trust legislation almost certainly played a critical rôle here, especially in persuading Bell Laboratories to put its patents into the public domain.) The UK offers neither: even after the mergers of the 1960s its larger firms were not big enough to compete with the established American giants, which had early penetrated British markets, thus also inhibiting the development of smaller specialized companies. Additionally, these American firms had developed sophisticated worldwide management systems, especially in market forecasting and monitoring and in feedback from the marketing to the design side (OECD 1968, 1969, Hays *et al.* 1969, Sciberras 1977, US National Research Council 1984, Stopford &Turner 1986).

The new rôle of the state

In the heroic age of American capitalist innovation, the 1880s and 1890s, none of it, with few and minor exceptions – as in the rôle of the US Census in aiding the rise of the Hollerith company, perhaps the critically important root of the IBM tree – owed anything to government action. It was first around the time of World War I, when first Germany (Telefunken, 1907) and then the USA (RCA, 1919) took the lead in creating radio oligopolies, that the state intervened directly for strategic reasons. But thereafter, and especially during and after World War II, the links between industry and government became extraordinarily strong and pervasive. From the start of the fourth Kondratieff the state everywhere played a major and active rôle in several different ways to promote New IT innovation and industrial growth: not merely through direct industrial-promotion policies but also defence policies, procurement policies, trading and tariff policies, education and training policies (Freeman *et al.* 1965, Rosenberg 1982).

In particular, as the USA moved out of isolationism and assumed the mantle of a great power, so the defence agencies began to develop direct contracting relationships with the private sector. Though never stated explicitly, military procurement became a highly effective programme of innovation-led economic development. Now, for the first time, the rôle of the state was no longer restricted to regulation, trade policy and educational provision; the American state itself began to invest on a huge scale in the contracting of R&D.

For, by definition, there could be only one dominant military power in the West. Perhaps the European Community nations could have established equal status, if they had consciously sought to collaborate in military procurement; but they did not do so. The result is most poignantly illustrated by the case of the UK, which vainly sought to retain great power status long after she had lost it, resulting in an extraordinary bias of research and development towards a military effort that was nevertheless doomed to ineffectuality. The paradox of the British failure, as we have told it in Chapter 12, is that its governments managed to get it wrong every way: wedded to 19th-century *laissez-faire* models, they failed to intervene to promote leading-edge New IT innovation by industry, on the Japanese model; instead, a different department of government followed the American path of defence-led innovation, but necessarily on a scale that was both too great for the country's straitened national resources and too small to achieve results. The result was bad economically, bad industrially, and bad militarily: an almost astonishingly inept achievement.

Defence–led innovation in IT

In any event, the real question is whether defence expenditure has truly proved important in aiding innovation, especially basic innovations that may effectively help create new industrial products and traditions. The balance of opinion, as set out in Chapter 12, is that at a critical historical juncture in the USA immediately after World War II it did so, speeding the development of the transistor and then the integrated circuit, and thus giving the USA a massive head lead in the whole complex of industries based on microelectronics. Yet, as just noticed, in the UK – a country that appeared excellently placed to compete with the USA for world leadership – a concentration on defence–led technology proved disastrous for the commercial competitiveness of the whole industry. And latterly, in the USA also, a powerful criticism has emerged that the same tendencies are evident (Stowsky 1986a, 1986b).

Stowsky's argument is that the requirements of defence contracting may run directly counter to the qualities needed for international competitiveness. Much defence–based R&D does not result in radical innovations; many contracts are for small quantities of specialized, non–standard products, thus inhibiting a tradition of cheap mass production; restrictions on export or publication deter commercial diffusion; defence companies tend to neglect efficiency and profit maximization, concentrating instead on political influence to win contracts. Thus, Stowsky argues, 'the increased importance of Pentagon contracts to the financial health of US high–tech firms threatens to undermine, from within, the nation's competitive position in a range of critical high–tech sectors' (Stowsky 1986b, 718).

This, he stresses, is not inevitable or invariable; Pentagon procurement policies did indeed boost the American competitive lead during the 1950s and 1960s. During the early years of integrated circuits, for instance, the Apollo moon programme and the Pentagon's Minuteman ICBM programme both required large-volume production. To meet this need, the semiconductor industry developed efficient production methods, which allowed it to drive down costs in the commercial sector. But, since 1970, the defence agencies have increasingly demanded complex specialized chips that are no longer in line with the needs of the commercial industry, which needs large volumes of cheap random–access memory chips to go into personal computers or video games. In this area, Japan has now begun to dominate through superb quality control and the repetition of standard circuit designs.

The paradox is thus that the last great defence spending boost, in the 1950s and 1960s, had a positive effect on the commercial development of the whole American NIT complex; the latest, in the 1980s, is having exactly the reverse effect. This could be fatal to the American competitive position.

The significance of societal innovation

The essential point, however, is both wider and more subtle: it is that really radical innovations, in the sense recognized by Schumpeter, occur in clusters, thus creating a new Technology System and a new 'Technico-Economic Paradigm' which embodies new systems of production, new kinds of industrial organization, new labour relationships and new locations of production. Thus the problem, central to this or any such study, is to unravel the precise sequence of forces that triggers such a new paradigm.

By definition, R&D leading to innovation, whether inside firms or in the wider scientific–educational milieu, must play a crucial rôle on the supply side; demand for the early products of innovation, whether from the market or from state agencies, must play a similar rôle on the other side. The relationship between these two is critical: successful R&D is a necessary but not a sufficient condition for successful major innovation, for there must also be rapid diffusion and adaptation of the resulting technologies among user-consumers (whether other companies, or individuals, or government) throughout the economy. These, too, must be prepared to innovate and to invest in innovation, to adopt the new technologies, and to change their organizational structures and methods accordingly; otherwise, the new products will lack home markets, producers will lack immediate user feedback, and without these as foundation they are unlikely even to begin to penetrate export markets. This is more than a matter simply of absorption of innovative products by existing consumers; often it involves the creation of entirely new sectors of the economy, such as radio and then television broadcasting, advanced telecommunications, and data processing. This, in turn, will involve a whole series of social, organizational and institutional innovations; the rate at which these occur will powerfully affect the rate of diffusion of the new technology system. So the country that makes the adaptation most rapidly will be the best placed to exploit the new technologies and to develop a world lead in their production.

This if anything makes it more difficult to understand the central paradox which emerged in Chapter 2: why radical innovatory clusters should be concentrated in Kondratieff downswings. For both supply-side R&D and demand-side customer innovation would logically be at a minimum then. Some observers have sought to explain this by exogenous forces, in particular state action – as in the rearmament wave of the mid-1930s. If this is so, the explanation for different Kondratieff recoveries may be quite diverse. Our analysis suggests that the innovatory boom of the 1870s and 1880s, and that of the 1930s and 1940s, had quite different origins. The first was commercially generated, and was prompted by the rapid secular growth of the American and German economies at the end of major wars; the second, as already seen, was in part state generated, and was prompted by the threat and then the reality of global conflict.

It is of course always possible that the world economy would have taken off anyway in the late 1940s and early 1950s, solely on the basis of market-technology electronic innovations like television and of revived older third Kondratieff technologies like cars or electrical goods; for it seems clear that much of the initial fourth Kondratieff expansion of the 1950s came from these sources, both directly and through their social-innovation accompaniments (suburbanization, travel, roadside culture). It is also possible that the key microelectronics innovations would have occurred in any case: the Bell system had a major interest in them, apparently without government support, from the mid-1930s, and Bell R&D was less affected than most by depression cutbacks. But without the massive postwar military funding of R&D they would almost certainly have come more slowly, and in particular the microelectronics technology chain would have been much more attenuated. Yet, as Stowsky implies, the same may not be true at all the next time around.

Innovatory cycles also vary between countries, and within countries, over time. The evidence strongly suggests that it is not sufficient for a country (or region, or city) to have an R&D lead in the development of new technologies: it must also have the right socio-economic milieu for rapid take-up of those technologies. The essential ingredients are difficult to define, but they include a pervasive ideology of scientific and technological progress; a high level of both general and technological education; a general feeling of commercial confidence in the future; an ability to develop large-scale, coherent, technological capacities, with supporting standards and regulatory policies; integrated corporate structures where needed; a capacity to develop and to diffuse new technologies; a high level of industrial and commercial investment; and access to long-term risk capital.

All these conditions prevailed in both Germany and the USA in the 1870s and 1880s, apparently overcoming the general worldwide depression of those decades; they were also true of the USA, and to a lesser extent Europe, in the recovery from World War II. They are thus particularly associated with economies enjoying fast-expanding internal markets, as was equally true of Germany after the unification of 1871; the USA during the last two frontier decades; and Western Europe during the period of the baby boom and the creation of the European economic communities. Such historical conditions appear crucial for the development of an innovatory managerial culture, in which marketing and servicing information is constantly fed back into the improvement and development of the product, producing a virtuous circle of innovation-led development (Freeman 1985). They also encourage the development of venture and finance capital, and increase the propensity of entrepreneurs to invest in high-risk developments, as well seen in the USA of the 1870s and the 1960s.

As Chapter 12 has stressed, Europe has lost these qualities since about 1960. There has been a relative failure both on the part of customers, in demand for the new products, and on the part of suppliers, in the ability to anticipate market trends; and these failures have interacted upon one another to produce a spiral of decline. Mackintosh's study suggests that unless a massive effort at renaissance is undertaken, by the year 2000 Europe as a whole will have virtually dropped out of the IT race, with a per capita IT production and a percentage IT employment lower than Japan's, with fewer IT jobs than today, and with a massive trade deficit in IT products (Mackintosh 1986).

But in the USA, also, by the late 1980s there are forebodings for the future. Japan is progressively increasing its hold on the mass production of high–quality, low–cost chips, thus pitting two major segments of the American IT industry in direct conflict: the chip fabricators demand protection against Japanese imports (arguing national security considerations in support), but the chip users want no such thing. The personal computer boom of the early 1980s appears to be largely spent, and production of low–cost IBM 'clones' is almost exclusively occurring offshore in low–cost areas like Taiwan and Korea. The deregulation of telecommunications has failed to give AT&T its expected break into computing, to compensate for the loss of its historic telecommunications monopoly; its UNIX–based personal computer has not proved a commercial success. Many American experts are concerned that the massive Japanese investment in the fifth generation computer will soon pay off, and that Japanese software writers will overcome the obvious difficulties they have experienced in breaking into the English–language market, thus giving that country a head lead in the 1990s.

The resulting fear, expressed openly in the American media, is that the USA will follow the British road to industrial atrophy. In April 1987 *Business Week* published an important special supplement headlined 'Can America compete?' It documented the worrisome long–term trends: a sharp decline in the rate of growth of GDP, an underlying growth rate of productivity of only 1% a year, a 1986 trade deficit of $170 billion, an external debt predicted to reach $750 million by 1990. It concluded, *inter alia*, that one of the country's main problems was its loss of export share in traditional industries (steel, machine tools, automobiles and consumer electronics) coupled with a failure to compete in the newer innovation–based high-technology products. Most ominously of all, perhaps, high-technology electronics products had shown a negative trade balance since 1984; in 1986 it was $13.1 billion, 49% greater than the year before.

The failure, *Business Week* stressed, lay not in lack of technological innovation; it lay in the inability of American industry to exploit that innovation. 'The Japanese', it concluded, 'created a manufacturing infrastructure that can respond with blazing speed to market demands and changing opportunities' (Port 1987, 57). The key was flexible production systems, coupled with a very high degree of automation and computer control of both design and production, yielding a wide range of tailor–made products that were both cheaper and better. American industry was beginning to compete by spending more on R&D, and American government was beginning to back that by spending more on basic scientific research – though most of that was on defence, where, as already noticed, it might not contribute much to economic regeneration.

The *Business Week* analysis ended with a telling historical comparison between American and British economic history. 'By and large', it concluded, 'Britain simply ceded industrial innovation to the US, and what it had, it used badly' (Pennar 1987, 66); as stressed also in our study, it failed to reorganize its 'atomistic industrial structure' into large corporations, capable of technological innovation. The USA, in contrast, had remained highly innovative; but perhaps the innovation was now being wasted, because of what the economist Mancur Olson called 'institutional sclerosis'. The USA, it appeared, was now making the same collective mistake as the UK had made from the 1880s: it was failing to utilize the fruits of its scientific lead. So, the supplement concluded, the USA might regain its lost world leadership; but that would require a whole set of major interlinked policy initiatives, among the most important of which would be an increase in R&D targeted to product and process development in key leading–edge technologies, such as biotechnology and superconductivity.

Towards the fifth Kondratieff

Most readers of this book are likely to be European, especially British, or American. Against the sombre background of UK and, latterly, European overall economic performance in the fourth Kondratieff, against the new worries for the US economy, they are bound to end by asking whether New IT can play a rôle in lifting these advanced industrial economies out of their late-1980s doldrums.

In Chapter 12 we concluded that, globally, the hypothesis of the SPRU group was disproved: even in the Kondratieff downswing, the New IT industries have provided plenty of new jobs. This, we hypothesize, is because really significant technological clusters – the carriers of their particular Kondratieffs – can continue to evolve during maturity, producing further technological innovations that are the source of significant job growth. This is well illustrated by the telephone (as an outgrowth of the telegraph) in the 1880s and by radio in the 1930s. We can indeed suggest that such technology clusters have a life cycle that may extend over two or more Kondratieffs, as well illustrated by the automobile industry. The emerging microelectronics-based New IT cluster, in particular, appears to have the historically rare quality that it can revolutionize the entire economic system: it has a wide range of possible applications ('pervasiveness'); it can bring great capital savings and thus profits; so it can increase the potential, and the expectation, for increased profits, setting up what Schumpeter called a 'secondary wave' (Schumpeter 1939; Soete & Dosi 1983; Perez 1985b, Soete 1985, Blackburn *et al.* 1985). This view is further reinforced by the remarkable breakthroughs, reported in the first half of 1987 as the final stage of this text was being completed, in the phenomenon of superconductivity of electricity at much higher temperatures than previously thought possible. For, according to early press reports, one of the most likely implications is a further dramatic increase in information–processing power and a commensurate reduction in cost (Gleick 1987, Lemonick 1987, Smith 1987).

The implications do not end there. For New IT can promote large increases in informational service jobs. In the UK, the proportion of such jobs increased from 7.7% to about 40% over the century from 1880 to 1980. In the USA, at the latter date, informational jobs accounted for over 44% of total employment (Hall 1987a, 1987b). From the early years of the third Kondratieff there was a notable shift towards new methods of production, one result of which was the growth of a large sector of administrative, managerial and clerical occupations within manufacturing. During the fourth there was a further growth in these white-collar sectors, to service the expansion of R&D, design, marketing, advertising and consumer credit, which became a distinct organizational innovation of that long wave. Until very recently productivity in these sectors was very low; in the late 1970s office workers had an average of only one-tenth of the capital equipment of shopfloor workers (Blackburn *et al.* 1985). Since then there is clear evidence of investment in systems such as word processors and advanced telephone systems. The proliferation of interactive systems, such as electronic mail and computer information systems, suggests that over the next decade investment may be far greater.

But some reservations need making. A number of American experts argue that the notion of a 'post-industrial' economy, based on 'the march through the sectors', is misleading. Though the American economy created a net 38.5 million new jobs between 1965 and 1986, all in the service sector, many of these jobs are low–skill and pay low wages (Bluestone & Harrison 1987). For this there appear to be two main reasons. Firstly, many of the new jobs are not in the so–called information sector, but in personal–service sector areas like hairdressing, health care, retailing or fast food, where skills are traditionally low and productivity gains difficult to achieve. Secondly, even in the information sector, productivity growth has been very slow. A recent study by the US Department of Labor in fact suggests that, paradoxically, productivity growth in this sector has sharply *fallen* during the 1980s, just as new technology has been injected into it. The reasons are obscure, but three have been identified: managers and employees do not know how to use the equipment properly; supervisors have not adopted effective management techniques; and equipment is too often unreliable (Schneider 1987).

Thirdly, and more subtly, the transition to the service economy may be a statistical illusion, because job growth in the service sector may really derive from the demands of the manufacturing sector. As Jonathan Gershuny in the UK and Stephen Cohen and John Zysman in the USA argue, if the input–output linkages are traced through, many jobs in the service economy in wholesale and retail trade, in transportation, and in banking, accountancy, law and other business services are dependent on manufacturing. Cohen and Zysman stress that no firm estimates exist, but hazard that in the USA, about half of GNP and employment is dependent on manufacturing (Cohen & Zysman 1987). Thus, they conclude:

> There is no such thing as a post–industrial economy. Manufacturing matters. The wealth and power of the United States depends on maintaining mastery and control of production. (ibid., 261)

Finally, there is an argument that service growth in the advanced countries, especially in the USA, may simply reflect lack of competition. Anecdotal evidence suggests for instance that American banks, American television and American advertising are very poorly managed and very lacking in innovation compared with their European equivalents (Kaletsky & de Jonquieres 1987).

The USA's failure may be Europe's and in particular the UK's opportunity. Many authorities suggest large possibilities for radical innovations in the provision of health, education, training and other public services, as well as for financial services, electronic sales and entertainment. Here, given the UK's relatively strong established position in many of these sectors, the country's best prospects may lie (ITAP 1983). And, since history suggests that the strength of the IT supply industries appears to depend greatly on the local demand for their products and services – both nationally, and also regionally – this may be the best strategy also for building a strong IT manufacturing sector.

Particularly significant, here, may be the possibility of making a radical shift from the standardized mass-production mode of manufacturing that characterized both the third and fourth Kondratieffs into the new mode of 'flexible specialization' (Piore & Sabel 1984): now, computer-controlled 'intelligent' machinery can introduce great variety into the process, making possible a form of low-cost bespoke production. This greater flexibility will affect not merely the direct production process but also the whole range of design, co-ordination, distribution and marketing. This, in turn, may well lead to a dramatic increase in the rôles of R&D, design, marketing and related work at the expense of manufacturing itself. Such functions may become the commanding economic activities of the fifth Kondratieff, with much of the actual manufacturing reduced to a subordinate rôle. Notable, here, is the increasing tendency of electronics 'manufacturers' (e.g. Amstrad) to concentrate on product design and marketing functions and to buy in the product itself from low-cost overseas suppliers.

But, as Cohen and Zysman argue, the race will go to the countries that themselves make the shift to the new production processes. Countries that simply rely on 'hollow corporations' to buy products that are manufactured offshore will end up losing the capacity to compete. For the importance of the new flexible systems does not merely lie in replacing labour by machinery; it lies in the massive cheapening of the production process, coupled with the greater variety of product (Cohen & Zysman 1987).

The best resolution of this debate, and of the strategic policy options, may lie somewhere in the middle. For European countries with historically strong international trading functions, like the UK, traded services are likely to play a more important rôle in production and in exports than they will for the USA. Some marketed information services (producer services) would be a critical, though not the only, part of a new service–oriented economy; others would be tourism and leisure services, health care services and education services. Each of these would make use, in varying – sometimes large – degree of Convergent IT in the form of hardware, software and communications systems. Not much of this would be manufactured in the UK, though, if past patterns continue, most should be designed there.

It is this final service package that will generate employment, income and exports; its composition will consist of a bundle of information–technology products – some manufactured in the UK, some imported – and the services they render. So the aim should be to maximize value added in the UK, not to maximize the indigenous production. It is almost certain to make sense to import hardware that can be made more cheaply and as effectively elsewhere, and to inject native ingenuity into its design, marketing and use. Equally, manufacturing will continue to matter: for the country that fails to master the production of the new technology is unlikely for long to lead in its use. But the production is only a means to an end: in the coming fifth Kondratieff, disembodied information itself will be an important carrier. This will be the first Kondratieff of the informational age, and the palm in the race will go to the nation that first seizes the fact.

References

ACARD 1980. *Information technology*. London: HMSO.

Aglietta, M. 1979. *A theory of capitalist regulation: the U.S. experience*. London: New Left Books.

Aldcroft, D. H. (ed.) 1968. *The development of British industry and foreign competition 1875–1914*. London: Allen & Unwin.

Aldcroft, D. H. 1970. *The inter-war economy: Britain 1919–1939*. London: Batsford.

Allen, G. C. 1970. *British industries and their organisation*. London: Longman.

Allesch, J. 1984. Die neuen Gründerjahre in Berlin: Das Entstehen von jungen Unternehmen im Umfeld der TUB. In *Die Zukunft der Metropolen: Paris–London–New York–Berlin: Katalog zur Austellung*, K. Schwarze (ed). Band 1: *Aufsätze*, 407–423. Berlin: Technische Universität Berlin.

Altshuler, A., M. Anderson, D. Jones, D. Roos, & J. Womack 1984. *The future of the automobile: the report of MIT's International Automobile Program*. London: Allen & Unwin.

Andre, D. 1971. *Indikatoren des technischen Fortschrifts: Eine Analyse der Wirtschaftsentwicklung in Deutschland von 1850 bis 1913*. Weltwirtschatliche Studien, 16. Göttingen: Van den Hoeck & Ruprecht.

Anon. 1954 *75 Jahre Mix & Genest 1879–1954*. ?Stuttgart. Privately printed.

Anon. 1956 *50 Jahre AEG*. Berlin: Allgemeine Elektrizitäts–Gesellschaft.

Anon. 1985. Editorial. *Computing* 5 December.

Archer, G. L. 1938. *History of radio to 1926*. New York: The American Historical Society.

Baar, L. 1966. *Der Berliner Industrie in der industriellen Revolution (Veröffentlichungen des Instituts für Wirtschaftsgeschichte an der Hochschule für Ökonomie Berlin- Karlshorst)*. Berlin: Akademie-Verlag.

BEAMA (British Electrical and Allied Manufacturers Association) 1927. *Combines and trusts in the electrical industry: the position in Europe in 1927*. London: BEAMA.

BEAMA 1929. *The electrical industry in Great Britain: organisation, efficiency and world competitive position*. London: BEAMA.

Beeching, W. A. 1974. *Century of the typewriter*. London: Heinemann.

Bernal, J. D. 1939. *The social function of science*. London: Routledge & Keegan Paul.

Berry, W. T. 1958. Printing and related trades. In *A history of technology. Vol. V: The late nineteenth century*, c1850 to c1900, C. Singer, E. J. Holmyard, A. R. Hall & T. I. Williams (eds), 683–715. Oxford: Clarendon Press.

Birdsall, D. & C. Cipolla 1980. *The technology of man: a visual history*. London: Wildwood House.

Blackburn, P., R. Coombs & K. Green 1985. *Technology, economic growth and the labour process*. London: Macmillan.

Bluestone, B. & B. Harrison 1987. The grim truth about the job 'miracle'. *The New York Times* 1 February.

Board of Trade 1925. *The world market for electrical goods*. Report of the Committee on Industry and Trade. London: BEAMA.

Braun, E. 1980. From transistor to microprocessor. In *The microelectronics revolution*, T. Forester (ed.), 72–82. Oxford: Basil Blackwell.

Braun, E. & S. Macdonald 1982. *Revolution in miniature: the history and impact of semiconductor electronics*, 2nd edn. Cambridge: Cambridge University Press.

Brock, G. W. 1975. *The U.S. computer industry: a study of market power*. Cambridge, Mass.: Ballinger.

Brooks, J. 1976. *Telephone: the first hundred years*. New York: Harper & Row.

Bruce, R. V. 1973. *Bell: Alexander Graham Bell and the conquest of solitude*. Boston: Little, Brown.

Byatt, I. C. R. 1979. *The British electrical industry 1875–1914: the economic returns to a new technology*. Oxford: Clarendon Press.

Cabinet Office 1985. *Annual review of government funded research and development*. London: HMSO.

Cambridge Econometrics 1985. *Electrical engineering industries*. Cambridge: Cambridge Econometrics.

Canby, E. T. 1962. *A history of electricity*. London: Prentice Hall.

Castells, M. (ed.) 1985. *High technology, space and society*. Urban Affairs Annual Reviews, 28. Beverly Hills and London: Sage.

Chandler, A. D. 1962. *Strategy and structure: chapters in the history of the industrial enterprise*. Cambridge, Mass.: MIT Press.

Chandler, A. D. 1975. *The visible hand: the managerial revolution in American business*. Cambridge, Mass.: MIT Press.

Chandler, A. D. 1984. The emergence of managerial capitalism. *Business History Review* **58**, 473–503.

Chesnais, F. 1982. Schumpeterian recovery and the Schumpeterian perspective: some unsettled issues and alternative interpretations. In *Emerging technologies: consequences for economic growth, structural change and employment*, H. Giersch (ed.), 33–71. Tubingen: J. C. B. Mohr.

Chew, V. K. 1981. *Talking machines*. London: HMSO.

Clapham, J. H. 1938. *An economic history of modern Britain: machines and national rivalries, 1887–1914*. Cambridge: Cambridge University Press.

Clark, N. 1985. *The political economy of science and technology*. Oxford: Basil Blackwell.

Cohen, S. S. & J. Zysman, 1987. *Manufacturing matters: the myth of the post–industrial economy*. New York: Basic Books.

COIT (Committee on Industry and Trade) 1927a. *Factors in industrial efficiency – Part I*. London: HMSO.

COIT 1927b. *Survey of overseas markets*. London: HMSO.

COIT 1928a. *Survey of metal industries*. London: HMSO.

COIT 1928b. *Further factors in industrial and commercial efficiency*. London: HMSO.

COIT 1929. *Final report of the Committee on Industry and Trade*. London: HMSO.

Commission of the European Communities 1975. *A study of the evolution of concentration in the mechanical engineering sector in the UK*. Brussels: European Commission.

Commission of the European Communities 1981. *A study of the evolution of concentration in the UK data processing industry with some international comparisons*. Luxembourg: Commission of the European Communities.

Committee on the Position of the Electrical Trades after the War 1918. Report. Cd.9072. London: HMSO.

Conot, R. 1979. *A stroke of luck*. New York: Seaview Books.

Corley, T. A. B. 1966. *Domestic electrical appliances*. London: Jonathan Cape.

Court, W. H. B. (ed.) 1965. *British economic history, 1870–1914: commentary and documents*. Cambridge: Cambridge University Press.

Crawley, H. J. 1957. *The National Research Development Corporation computer project*. NRDC Computer Sub-Committee paper 132, London. In *The story of vintage computers and the people who built them*, S. Lavington, 1980. Manchester: Manchester University Press.

CSO (Central Statistical Office) 1985. The UK balance of payments. *Economic Trends* March, 72-91.

Czada, P. 1969. *Der Berliner Elektroindustrie in der Weimarer Zeit: Eine regionalstatistische-wirtschaftliche Untersuchung*. Berlin: Colloquium.

Dorfman, N. S. 1982. *Massachusetts' high technology boom in perspective: an investigation of its dimensions, causes and of the role of new Firms*. Cambridge, Mass.: MIT, Center for Policy Alternatives (CPA 82-2).

Dorfman, N. S. 1983. Route 128: the development of a regional high technology economy. *Research Policy* **12**, 299-316.

Dosi, G. 1984a. *Technical change and industrial transformation: the theory and an application to the semiconductor industry*. London: Macmillan.

Dosi, G. 1984b. Technological paradigms and technological trajectories: the determinants and directions of technological change and transformation of the economy. In *Long waves in the world economy*, C. Freeman (ed.), 78–101. London: Pinter.

Du Boff, R. B. 1980. Business demand and the development of the telegraph in the United States, 1844–1860. *Business History Review* **54**, 459–68.

Duckworth, G. H. 1895. Scientific, surgical and electrical instruments. In *Life and labour of the people in London: VI*, C. Booth, (ed.), 35–53. London: Macmillan.

Dummer, G. W. A. 1983. *Electronic inventions and discoveries*. Oxford: Pergamon.

EITB (Engineering Industry Training Board) 1984. *Manpower in the electronic industries*. Watford: EITB.

Elbaum, B. & W. Lazonick 1984. The decline of the British economy: an institutional perspective. *Journal of Economic History* **44**, 567–85.

Elbaum, B. & W. Lazonick, W. 1985. *The decline of the British economy*. Oxford: Oxford University Press.

Electronics Magazine 1981. *An age of innovation: the world of electronics 1930–2000*. New York: McGraw-Hill.

Ernst, D. 1985. Automation and the worldwide restructuring of the electronics industry: strategic implications for developing countries. *World Development* **13**, 333–52.

Estall, R. C. 1966. *New England: a study in industrial adjustment*. London: Bell.

Ewers, H.-J. 1984. Berlin: Auf dem Wege zur Industriemetropole der Zukunft? In *Die Zukunft der Metropolen: Paris-London-New York-Berlin: Katalog zur Austellung*, K. Scharze (ed). Band 1: *Aufsätze*, 397–406. Berlin: Technische Universität Berlin.

Fagen, M. D. (ed.) 1975. *A history of science and engineering in the Bell System: the early years (1975–1925)*. Murray Hill, NJ: Bell Telephone Laboratories.

Fagen, M. D. (ed.) 1978. *A history of science and engineering in the Bell System: national service in war and peace (1925–1975)*. Murray Hill, NJ: Bell Telephone Laboratories.

Finn, B. S. 1985. Telecommunications. *Encyclopedia Britannica* **28**, 495–9.

Fischer, W. 1976. Bergbau, Industrie und Handwerk. In *Handbuch der deutschen Wirtschafts- und Sozialgschichte*, H. Aubin & W. Zorn (eds.), 527–62. Stuttgart: Ernst Klett.

Fishman, K. D. 1981. *The computer establishment*. New York: McGraw-Hill.

Fleming, A. P. M. & H. J. Brocklehurst 1925. *A history of engineering*. London: Black.

Floud, R. & D. McCloskey (eds) 1981. *The economic history of Britain since 1700*. Vol. 2: *1860 to the 1970s*. Cambridge: Cambridge University Press.

Forester, T. (ed.) 1980. *The microelectronics revolution*. Oxford: Basil Blackwell.

Forester, T. (ed.) 1985. *The information technology revolution*. Oxford: Basil Blackwell.

Forester, T. 1987. *High–tech society*. Oxford: Basil Blackwell.

Freeman, C. 1962. Research and development: a comparison between British and American industry. *National Institute Economic Review* **20**, 21–39.

Freeman, C. 1978. *Technical innovation and British trade performance*. In *Deindustrialisation*, F. Blackaby (ed.), 56–73. London: Heinemann.

Freeman, C. (ed.) 1984. *Long waves in the world economy*. London: Pinter.

Freeman, C. 1985. Long waves of economic development. In *The information technology revolution*, T. Forester (ed.), 602–16. Oxford: Basil Blackwell.

Freeman, C., C. J. F. Harlow & J. K. Fuller 1965. Research and development in electronic capital goods. *National Institute Economic Review* **34**, 40–88.

Freeman C., J. Clark & L. Soete 1982. *Unemployment and technical innovation – a study of long waves and economic development*. London: Pinter.

Freiberger, P. & M. Swaine 1984. *Fire in the valley: the making of the personal computer*. Berkeley, Calif.: Osborne/McGraw Hill.

Garcke, E. 1896. *Manual of electrical undertakings: first year*. London: P. S. King & Son.
Garcke, E. 1907. *The progress of electrical enterprise*. London: Electrical Press Limited.
Gershuny, J. 1987. Lifestyles, innovation and the future of work. *Journal of the Royal Society of Arts* **135**, 492–502.
Gershuny, J. & I. Miles 1983. *The new service economy*. London: Pinter.
Gleick, J. 1987. Discoveries bring a 'Woodstock' for physics. *The New York Times* 20 March.
Guy, K. 1985. Communications. In *Electronics and communications*, L. Soete (ed.), 90–145. Aldershot: Gower.

Hall, P. G. 1962. *The industries of London since 1861*. London: Hutchinson.
Hall, P. 1984. Metropolis 1890–1940: challenges and responses. In *Metropolis 1890–1940*, A. Sutcliffe, (ed.), 19–66. London: Mansell.
Hall, P. 1985. The geography of the fifth Kondratieff. In *Silicon landscapes*, P. Hall & A. Markusen (eds), 1–19. Boston: Allen & Unwin.
Hall, P. 1987. The anatomy of job creation: nations, regions and cities in the 1960s and 1970s. *Regional Studies* **21**, 95–106.
Hall, P. 1988. Regions in the transition to the information economy. In *America's new economic geography*, G. Sternlieb (ed.), New Brunswick: Rutgers University, Center for Urban Policy Research. Forthcoming.
Hall, P., M. Breheny, R. McQuaid, & D. Hart 1987. *Western sunrise: the genesis and growth of Britain's major high tech corridor*. London: Allen & Unwin.
Handel, S. 1967. *The electronic revolution*. London: Penguin.
Hannah, L. 1979. *Electricity before nationalisation*. London: Macmillan.
Harman, A. J. 1971. *The international computer industry: innovation and comparative advantage*. Cambridge, Mass.: Harvard University Press.
Hays, S., M. F. Hemming, & G. F. Ray 1969. The office machinery industry in the United Kingdom. *National Institute Economic Review* **49**, 52–74.
Hegemann, W. 1930. *Das steinerne Berlin: Geschichte der grössten Mietskasernenstadt der Welt*. Berlin: Gustav Kiepenheues.
Heims, S. J. 1980. *John von Neumann and Norbert Wiener: from mathematics to the technologies of life and death*. Cambridge, Mass.: MIT Press.
Henderson, W. O. 1958. *The state and the industrial revolution in Prussia 1740–1870*. Liverpool: Liverpool University Press.
Henderson, W. O. 1975. *The rise of German industrial power 1834–1914*. London: Temple Smith.
Henning, F.-W. 1974. *Das industrialisierte Deutschland 1914 bis 1972*. Paderborn: Ferdinand Schöningh.
Hewer, A. 1980. Manufacturing industry in the 1970s: an assessment of import penetration and export performance. *Economic Trends* **326**, 97–109.
Hewlett Packard Ltd 1985. *Memorandum to House of Lords Select Committee on Overseas Trade*. Report, Vol. 2, 479–486. London: HMSO.
Hills, J. 1984. *Information technology and industrial policy*. London: Croom Helm.
Hjulstrom, F. 1942. *An economic geography of electricity: an outline*. Uppsala: University, Geographical Institute.

Hobsbawm, E. J. 1969. *Industry and empire*. London: Penguin.

Hoffman, K. 1985. Microelectronics. *In* International competition and development strategies. *World Development* **13**, 262–272.

Holthaus, E. 1980. *Die Entwicklung der Produktivkräfte im Deutschland nach der Reichsgründung bis zur Jahrhundertwende*. Frankfurt: Haag & Herchen.

Hughes, T. P. 1983. *Networks of power: electrification in western society, 1880–1930*. Baltimore, Md.: Johns Hopkins University Press.

Hund, J. M. 1959. Electronics. In *Made in New York: case studies in metropolitan manufacturing*, M. Hall (ed.), 241–308. Cambridge, Mass.: Harvard University Press.

Inose, H. & J. R. Pierce 1984. *Information technology and civilisation*. New York: Freeman.

ITAP (Information Technology Advisory Panel) 1983. *Making a business of information*. London: HMSO.

Jarvis, C. M. 1958. The distribution and utilization of electricity. In *A history of technology*. Vol. 5: *The late nineteenth century, c1850 to c1900*, C. Singer, E. J. Holmyard, A. R. Hall & T. I. Williams, 177–207. Oxford: Clarendon Press.

Jewkes, J., D. Sawers, & R. Stillerman 1967. *The sources of invention*, 2nd edn. London: Macmillan.

Johnson, J. H. & W. L. Randell 1945. *Colonel Crompton and the evolution of the electrical industry*. London: The British Council.

Jones, R. & O. Marriott 1970. *Anatomy of a merger: a history of GEC, AEI and English Electric*. London: Jonathan Cape.

Josephson, M. 1961. *Edison: a biography*. London: Eyre & Spotiswoode.

Josephson, M. 1985. Edison. *Encyclopaedia Britannica* **17**, 1049–51.

Kaletsky, A., & G. de Jonquieres 1987. Why a service economy is no panacea. *Financial Times*, 22 May.

Katz, B. G. & A. Phillips 1982. Government, technological opportunities and the emergence of the computer industry. In *Emerging technologies: consequences for economic growth, structural change and employment*, H. Giersch (ed.), 419–66. Tübingen: J. C. B. Mohr.

Keeble, D. E. 1968. Industrial decentralization and the metropolis: the North-West London case. *Transactions and Papers, Institute of British Geographers* **44**, 1–54.

Kindleberger, C. P. 1978. *Economic response – comparative studies in trade, finance and growth*. Cambridge, Mass.: Harvard University Press.

Kirby, M. W. 1981. *The decline of British economic power since 1870*. London: Allen & Unwin.

Kleinknecht, A. 1984. Observations on the Schumpeterian swarming of innovations. In *Long waves in the world economy*, C. Freeman (ed.), 48–62. London: Pinter.

Kocka, J. 1969. *Unternehmensverwaltung und Angestellenschaft am Beispiel Siemens 1847–1914: zum Verhältnis von Kapitalismus und Buřokratie in der deutschen Industrialisierung*. Stuttgart: Ernst Klett.

Kondratieff, N. D. 1926. Die langen Wellen der konjunktur. *Archiv für Sozialwissenschaft*, **56**, 573–609.

Kondratieff, N. D. 1935. The long waves in economic life. *The Review of Economic Statistics* **17**, 105–15.
Kondratieff, N. D. 1984. The long wave cycle. New York: Richardson and Snyder.
Kuznets, S. S. 1940. Schumpeter's business cycles. *American Economic Review* **30**, 250–71.

Landes, D. S. 1969. *The unbound Prometheus*. Cambridge: Cambridge University Press.
Lemonick, M. D. 1987. Superconductors! *Time* 11 May, 64–75.
Loria, J. 1984. Die Massachusetts Institute of Technology und die Entwicklung der Region Boston. In *Die Zukunft der Metropolen: Paris–London–New York–Berlin: Katlog zur Ausstellung*, K. Schwarz (ed.), Band 1, Aufsähe 129–46. Berlin: Technische Universität.

Mackintosh, I. 1986 *Sunrise Europe: the dynamics of information technology*. Oxford: Basil Blackwell.
MacLaren, M. 1943. *The rise of the electrical industry during the nineteenth century*. Princeton, NJ: Princeton University Press.
Maclaurin, W. R. 1949. *Invention and innovation in the radio industry*. New York: Macmillan.
MacLeod, R. & K. MacLeod 1977. The social relations of science and technology 1914–39. In *The twentieth century*, C. M. Cipolla (ed.), 301–35. Fontana Economic History. London: Fontana.
Mandel, E. 1975. *Late capitalism*. London: New Left Books.
Mandel, E. 1980. *Long waves of capitalist development*. Cambridge: Cambridge University Press.
Manegold, K.-H. 1970. *Universität, Technische Hochschule und Industrie: Ein Beitrag zur Empanzipation der Technik im 19. Jahrhundert unter besondere Berücksichtigung des Bestrebungen Felix Klein*. Berlin: Duncker & Humblot.
Marchetti, C. 1985. Swings, cycles and the global economy. *New Scientist*, 2 May, 12–15.
Marchetti, C. 1986. Fifty-year pulsation in human affairs: analysis of some physical indicators. *Futures* **18**, 376–88.
Margerison, T. A. 1978. Computers. In *A history of technology*. Vol. VI: *The twentieth century c1900 to c1950*, T. I. Williams (ed.), 1150-1203. Oxford: Clarendon Press.
Markusen, A., P. Hall & A. Glasmeier 1986. *High-tech America: the what, how, where and why of the sunrise industries*. Boston: Allen & Unwin.
Martin, J. E. 1966. *Greater London: an industrial geography*. London: Bell.
Mensch, G. 1979. *Stalemate in technology*. Cambridge, Mass.: Ballinger.
Metropolis, N., J. Howlett & G.-C. Rota (eds) 1980. *A history of computing in the twentieth century: a collection of essays*. New York: Academic Press.
Moran, J. 1978. Printing. In *A history of technology*. Vol. VI: *The twentieth century c1900 to c1950*, T. Williams (ed.), 1268–80. Oxford: Clarendon Press.
Mowery, D. C. 1983. Economic theory and government technology policy. *Policy Sciences* **16**, 27–43.

Mowery, D. C. 1984. Firm structure, government policy and the organisation of industrial research: Britain and the US 1900–1975. *Business History Review* **58**, 504–31.

National Research Council 1984. *The competitive status of the US electronics industry*. Washington DC: National Academy Press.

National Science Board 1983. *Science Indicators 1983*. Washington: Government Printing Office.

NEDO (National Economic Development Office) 1967. *Statistics of the electronics industry*. London: NEDO.

NEDO 1968. *Electronics industry statistics and their sources*. London: NEDO.

NEDO 1970. *Industrial report by the electronics EDC: On the economic assessment to 1972*. London: NEDO.

NEDO 1984. *Crisis facing the UK's information technology industry*. London: NEDO.

Noble, D. F. 1977. *America by design: science, technology, and the rise of corporate capitalism*. New York: Knopf.

OECD (Organisation for Economic Cooperation and Development) 1968. *Gaps in technology: electronic components*. Paris: OECD.

OECD 1969. *Gaps in technology: electronic computers*. Paris: OECD.

OECD 1970. *Gaps in technology: education, research and development, technological innovation in international economic exchanges*. Paris: OECD.

OECD 1981. *Information activities, electronics, and telecommunications technologies*. Information, Computer and Communications Policy No. 6. Paris: OECD. 2 volumes.

OECD 1985a. *Software: an emerging industry*. Paris: OECD.

OECD 1985b. *The semiconductor industry–trade related issues*. Paris: OECD.

Olson, M. 1982. *The rise and decline of nations: economic growth, stagflation and social rigidities*. New Haven: Yale University Press.

Oules, F. 1966. *Economic planning & democracy*. London: Penguin.

Passer, H. C. 1952. Electrical manufacturing around 1900. *Journal of Economic History* **12**, 378–95.

Passer, H. C. 1953. *The electrical manufacturers 1875–1900: a study in competition, entrepreneurship, technical change, and economic growth*. Cambridge, Mass.: Harvard University Press.

Pavitt, K. (ed.) 1980. *Technical innovation and British economic performance*. London: Macmillan.

Payne, G. L. 1960. *Britain's scientific and technical manpower*. Cambridge: Cambridge University Press.

Peck, M. J. 1968. Science and technology. In *Britain's economic prospects*, R. E. Caves (ed.), 448–84. London: Allen & Unwin.

Peck, M. J. & R. W. Wilson 1982. Innovation, imitation, and comparative advantage: the performance of Japanese color television set producers in the U.S. market. In *Emerging technologies: consequences for economic growth, structural change and employment*, H. Giersch (ed.), 195–212. Tubingen: J. C. B. Mohr.

Pennar, K. 1987. Is the U.S. going the way of Britain? *Business Week* 20 April, 64–66.
Perez, C. 1983. Structural change and the assimilation of new technologies in the economic and social systems. *Futures* **15**, 357–75.
Perez, C. 1985. Microelectronics, long waves and world structural changes: new perspectives for developing countries. *World Development* **13**, 441–63.
Perroux, F. 1961. *L'Economie du XXième Siecle*. Paris: Presses Universitaires de France.
Piore, M. J. & C. F. Sabel 1984. *The second industrial divide: possibilities for prosperity*. New York: Basic Books.
Pollard, S. & C. Holmes 1972. *Industrial power and national rivalry 1870–1914*. London: Edward Arnold.
Pope, R. & B. Hoyle 1985. *British economic performance 1880–1980*. London: Croom Helm.
Porat, M. 1977. *The information economy*. Washington: US Department of Commerce.
Port, O. 1987. Making brain work with brawns. *Business Week* 20 April, 56–60.
Prais, S. J. 1981. *Productivity and industrial structure*. Cambridge: Cambridge University Press.
Preston, P. 1987. Technology waves and the future sources of employment and wealth creation in Britain. In *The development of high technology industries: an international survey*, M. Breheny & R. McQuaid (eds). London: Croom Helm.
Preston, P., P. Hall, & N. Bevan 1985. *Innovation in information technology industries in Great Britain*. Reading: Reading University Geographical Papers No. 89.

Quigley, H. 1925. *Electrical power and national progress*. London: Allen & Unwin.

Randell, B. 1980. The COLOSSUS. In *A history of computing in the twentieth century: a collection of essays*, N. Metropolis, J. Howlett, & G.-C. Rota (eds), 47–92. New York: Academic Press.
Reich, L. 1985. *The making of American industrial research: science and business at G.E. and Bell, 1876–1926*. Cambridge: Cambridge University Press.
Richardson, H. W. 1967. *Economic recovery in Britain, 1932–9*. London: Weidenfeld & Nicolson.
Roderick, G. W. & M. D. Stephens 1972. *Scientific and technical education in nineteenth–century England*. Newton Abbot: David & Charles.
Rogers, E. M. & J. K. Larsen 1984. *Silicon Valley fever: growth of high-technology culture*. New York: Basic Books.
Rosenberg, N. (ed.) 1976. *Perspectives on technology*. Cambridge: Cambridge University Press.
Rosenberg, N. 1982. *Inside the black box: technology and economics*. Cambridge: Cambridge University Press.
Rosenberg, N. & C. Frischtak 1984. Technological innovation and long waves. *Cambridge Journal of Economics* **8**, 7–24.
Rothwell, R. 1982. The role of technology for industrial change: implications for regional policy. *Regional Studies* **16**, 361–9.

Rothwell, R. & W. Zegveld 1981. *Industrial innovation and public policy: preparing for the 1980s and 1990s*. London: Frances Pinter.
Rowland, J. 1960. *Progress in power*. London: Newman Neame.
Russon, A. R. 1985. Writing. *Encyclopaedia Britannica* **29**, 1008–10.

Sandberg, L. G. 1981. The entrepreneur and technological change. In *The economic history of Britain since 1700. Vol. 2: 1860 to the 1970s*, R. Floud & D. McCloskey (eds), 99–120. Cambridge: Cambridge University Press.
Sanderson, M. 1972a).*The universities and British industry, 1880–1970*. London: Routledge & Kegan Paul.
Sanderson, M. 1972b. Research and the firm in British industry 1919–1939. *Science Studies* **2**, 107–52.
Saul, S. B. 1960. The American impact on British industry 1895–1914. *Business History* **3**, 19–38.
Saxenian, A. 1985a. Innovative manufacturing industries: spatial incidence in the United States. In *High technology, space and society*, M. Castells (ed.), 55–80. Beverly Hills and London: Sage.
Saxenian, A. 1985b. The genesis of Silicon Valley. In *Silicon landscapes*, P. Hall & A. Markusen (eds), 20–34. Boston: Allen & Unwin.
Saxenian, A. L. 1985c. Let them eat chips. *Environment and Planning, D (Society and Space)* **3**, 121–7.
Schneider, K. 1987. High tech actually cuts productivity in U.S. service industry. *New York Herald Tribune* 29 June.
Schremmer, W. 1973. Wie gross war der technische Fortschrift während der Industriellen Revolution in Deutschland, 1856–1913. *Vierteljahrschrift für Sozial- und Wirtschaftsgeschichte* **60**, 433–58.
Schulz-Hanssen, K. 1970. *Die Stellung der Elektroindustrie im Industrialisierungsprozess (Schriftenreihe zur Industrie- und Entwicklungspolitik, Band 5)*. Berlin: Duncker & Humblot.
Schumpeter, J. A. 1928. The instability of capitalism. *Economic Journal* **38**, 361–86.
Schumpeter, J. A. 1939. *Business cycles*. New York: McGraw-Hill.
Schumpeter, J. A. 1942. *Capitalism, socialism and democracy*. New York: Harper.
Schumpeter, J. A. 1952. *Ten great economists: from Marx to Keynes*. London: Allen & Unwin.
Schumpeter, J. A. 1954. *History of economic analysis*. New York: Oxford University Press.
Schumpeter, J. A. 1961 (1911). *The theory of economic development*. Oxford: Oxford University Press.
Sciberras, E. 1977. *Multinational electronics companies and national economic policies*. Greenwich, Conn.: Jai Press.
Scott, A. J. 1986. High technology industry and territorial development: the rise of the Orange County complex, 1955–84. *Urban Geography* **7**, 3–45.
Scott, J. D. 1958. *Siemens Brothers 1858–1958: an essay in the history of industry*. London: Weidenfeld & Nicolson.
Shonfield, A. 1958. *British economic policy since the war*. London: Penguin.
Siemens, G. 1957a. *History of the House of Siemens*. Vol. I: *The era of free enterprise*. Freiburg and Munich: Karl Alber.

Siemens, G. 1957b. *History of the House of Siemens*. Vol. II: *The era of world wars*. Freiburg and Munich: Karl Alber.

Singelmann, J. 1978. *From agriculture to services*. Beverly Hills: Sage.

Singer, C., E. J. Holmyard, A. R. Hall & T. I. Williams 1958. *A history of technology*. Vol. V. *The late nineteenth century, c1850 to c1900*. Oxford: Clarendon Press.

Smith, D. H. 1933. *The industries of Greater London*. London: P. S. King.

Smith, E. T. 1987. 'Our Life has changed': the lightbulb, the transistor – now the superconductor revolution. *Business Week* 6 April, 94–7.

Smits, F. M. (ed.) 1985. *A history of science and engineering in the Bell system: electronics technology (1925–76)*. (?)Murray Hill, NJ: AT&T Bell Laboratories.

Sobel, R. 1983. *IBM: colossus in transition*. New York: Bantam Books.

Soete, L. 1985a. International diffusion of technology, industrial development and technological leapfrogging. *World Development* **13**, 409-22.

Soete, L. (ed.) 1985b. *Electronics and communications*. Aldershot: Gower.

Soete, L. & G. Dosi 1983. *Technology and employment in the electronics industry*. London: Frances Pinter.

Solomou, S. 1986. Innovation clusters and Kondratieff long waves in economic growth. *Cambridge Journal of Economics* **10**, 101–12.

Starr, H. E. 1944. *Dictionary of American biography*. Vol. 21. (Supplement 1). London: Oxford University Press.

Sterman, J. D. 1985. An integrated theory of the long wave. *Futures* **17**, 104–31.

Steward, F. & D. Wield 1984. Science planning and the state. In *The State and Society* (Block 4), Open University (ed.), Milton Keynes: Open University.

Stonier, T. 1983. *The wealth of information*. London: Methuen.

Stoneman, P. 1976. *Technological diffusion and the computer revolution: the UK experience*. Cambridge: Cambridge University Press.

Stopford, J. M. & L. Turner 1986. *Britain and the multinationals*. Chichester: Wiley.

Storbeck, D. 1964. *Berlin: Bestand und Möglichkeiten (Dortmunder Schriftren zur Sozialforschung, 27)*. Köln/Opladen: Westdeutscher Verlag.

Stowsky, J. 1986a. *Beating our plowshares into double–edged swords: the impact of Pentagon policies on the commercialization of advanced technologies*. BRIE Working Paper No. 17. Berkeley: University of California, Berkeley Roundtable on the International Economy.

Stowsky, J. 1986b. Competing with the Pentagon. *World Policy Journal* **3**, 697–721.

Strandh, S. 1979. *Machines: an illustrated history*. London: Artists House.

Sturmey, S. G. 1958. *The economic development of radio*. London: Duckworth.

Sutcliffe, A. (ed.) 1984. *Metropolis 1890–1940*. London: Mansell.

Taylor, H. G. 1970. *An experiment in co-operative research : an account of the first fifty years of the electricity research association*. London: Hutchinson.

Tillman, J. R. & D. G. Tucker 1978. Electronic engineering. In *A history of technology*. Vol. VI: *The twentieth century c1900 to c1950*, T. I. Williams (ed.), 1091–1125. Oxford: Clarendon Press.

Tilton, J. E. 1971. *International diffusion of technology: the case of semiconductors*. Washington, DC: The Brookings Institution.

Toffler, A. 1980. *The third wave*. London: Pan.

Toffler, A. 1983. *Previews and premises*. London: Pan.

Treue, W. 1970. Gesellschaft, Wirtschaft und Technik im 19. Jahrhundert. In *Handbuch der Deutschen Geschichte* 9th edn., B. Gebhardt, neu bearbetete Auflage herausgeben von H. Grundmann, **3**, 377–541. Stuttgart: Union.

Treue, W. 1976. Die Technik in Wirtschaft und Gesellschaft 1800–1970. In *Handbuch der deutschen Wirtschafts- und Sozialgschichte*, H. Aubin & W. Zorn (eds), 51–121. Stuttgart: Ernst Klett.

Tucker, D. G. 1978. Electrical communication. In *A history of technology. Vol. VI: The twentieth century c1900 to c1950*, T. I. Williams (ed.), 1220–67. Oxford: Clarendon Press.

Uselding, P. 1980. Business history and the history of technology. *Business History Review* **54**, 443–54.

US National Research Council 1984. *The competitive status of the US electronics industry*. Washington, DC: National Academy Press.

van Duijn, J. J. 1983. *The long wave in economic life*. London: Allen & Unwin.

Vernon, R. 1966. International investment and international trade in the product cycle. *Quarterly Journal of Economics*, **80**, 190–207.

von Peschke, H.-D. 1981. *Elektroindustrie und Staatsverwaltung am Beispiel Siemens 1847–1914*. Frankfurt-am-Main und Bern: Peter D. Lang.

von Weiher, S. 1974. *Berlins Weg zur Elektropolis*. Berlin: Stapp.

von Weiher, S. & H. Goetzeler, 1981. *Weg und Wirken der Siemens-Werke im Fortschritt der Elektrotechnik 1847–1980: Ein Beitrag zur geschichte der Elektroindustrie (Zeitschift für Unternehmensgeschichte, Beiheft 21)*. Wiesbaden: Franz Steiner.

von Winterfeld, L. 1913. *Entwicklung und Täkigkeit der Firma Siemens und Halkse in den Jahren 1847–92*. ? Berlin: ? privately printed.

Wachhorst, W. 1981. *Thomas Alva Edison: an American myth*. Cambridge, Mass.: MIT Press.

Walker, W. B. 1980. Britain's industrial performance 1850–1950: a failure to adjust. In *Technical innovation and British economic performance*, K. Pavitt (ed.), 19–37. London: Macmillan.

Wiener, M. 1981. *English culture and the decline of the industrial spirit*. Cambridge: Cambridge University Press.

Wildes, K. L., & N. A. Lindgren 1985. *A century of electrical engineering and computer science at M.I.T., 1882–1982*. Cambridge, Mass.: MIT Press.

Williams, T. I. (ed.) 1978. *A history of technology*. Vol. VI. *The twentieth century c1900 to c1950*. Part II. Oxford: Clarendon Press.

Wilson, T. 1964. The electronics industry. In *The structure of British industry*, D. Burn (ed.), 130–83. Cambridge: Cambridge University Press.

Wise, G. 1985. *Willis R. Whitney, General Electric, and the origins of U.S. industrial research*. New York: Columbia University Press.

Wise, M. J. 1949. On the evolution of the jewellery and gun quarters in Birmingham. *Institute of British Geographers, Transactions and Papers* **15**, 57–72.

Zarnowitz, V. 1985. Recent work on business cycles in historical perspective: a review of theories and evidence. *Journal of Economic Literature* **23**, 523–80.

Zimm, A. 1959. *Die Entwicklung des Industriestandortes Berlin: Tendenzen der geographischen Lokalisation bei der Berliner Industriezweigen von überortliche Bedeutung sowie die territoriale Stadtentwicklung bis 1945.* Berlin (E.): Deutscher Verlag der Wissenschaften.

Zuse, K. 1980. Some remarks on the history of computing in Germany. In *A history of computing in the twentieth century: a collection of essays*, N. Metropolis, J. Howlett & C.-G. Rota (eds), 611–27. New York: Academic Press.

Index

Illustrations are indicated by italicised figure numbers.

a.c. power system, 63–5
AEG 127, Table 8.2, 128, 129, 136, 137
  locations 131
American Marconi Company 92, 142
Apple 159
AT & T 91, 119, 142, 283

Baird, John Logie 96–7
batteries, electric 39, 58, 62–3
BBC 95, 96–7
Bell, Alexander Graham 48–9, 140–1
Bell Laboratories 141, 152, 160
Berlin 124–37, 243–4
Berliner, Emil 80–1
Boston-New York corridor 137–45, 254, *see also* Highway *128*
British Insulated Wire Company 47
British Marconi Company 89, 90, 92, 107
British Tabulating Machine Company (BTM) 225
broadcasting, radio 94–5
  equipment 96, 206–9, 227–8
Brush, Charles, F. 60–1, 68
BTM 225
business enterprises *see* industrial organizations

cable industry 45–7, 146
calculators 83–4
  electronic 164
capital goods, electronic
  classification 173–4
  see also under names of individual goods
cash registers 171
cathode ray tubes 97, Table 10.6, 229
Chicago 142, 143
CIT 5, 274
components, electronic 181–5
  employment, industrial 197–8, *12.2*
  production 184–5, 228–30
  trade, international 183, 209–10, Table 12.2
  see also under names of individual components
computers 154–9, 175–7, 224–7, 283
consumer goods, electronics 162–3, 197–8, 223–4
Convergent Information Technology 5, 274

Davy, *Sir* Humphrey 67
d.c. generators 59–63
defence spending 174, 182, 236–41, 279–80
Department of Scientific and Industrial Research 118–19
dictation machines 79–80, 171
digital transmission 160–1, 162
diode valves 88, 91
duplicating machines
  electrostatic 164, 171–2
  stencil 83, 171

Edison, Thomas Alva 41, 61, 78, 83, 137–40
  and telephone system 49, 51
Edison effect 87–8
education, scientific and technical 110–12, 116–17
EEC 213
  employment, electronics 193, 196, 197, 198–9, *11.2*
  trade, international 204–5, 206–7, 210, 211
  see also Germany; United Kingdom
electricity, early discoveries 38–9
electrostatic copying 164, 171–2
Elektropolis (Berlin) 124–37, 243–4
EMI 97
employment, industrial
  electronics 191–202
  informational services 285–6
  research 99
  women in offices 77
  see also under names of individual countries
English Telegraph Company 42
exchanges, telephone 50, 160

Fairchild Semiconductor Ltd 152, 153, 257
Ferranti, Sebastian Ziani de 47, 66
Ferranti PLC 225, 227
Fessenden, Reginald 90–1
finance capital 112, 121, 135, 275
Fleming, Ambrose 88
Forest, Lee D. 88

General Electric Company 61–2, 114, 140, 141, 142
Germany 52, 92–3, 117
  employment, industrial *11.2*, Table 11.2, Table 11.4
    electrical Table 7.5, 104
    electronics *11.4*, *11.5*
    office machinery *11.3*
  patents 232, Table 12.9, Table 12.10
  production 111
    electrical goods 102
    electronic goods 179, 180, Tables 10.4–10.6
    office machinery 167
  research and development Table 12.11–12.14, Table 12.15
  trade, international 204, Table 11.7
    electrical goods Table 7.4
    electronic goods Table 10.5, *11.11 – 11.14*
    office machinery 169, Table 11.6, *11.10*
  see also Berlin
governments 122, 181–2, 278–9, *see also* defence spending; nationalization
Gramme, Zenobe Theophile 60
gramophone 80
Gray, Elisha 48
Great Britain *see* United Kingdom

Henry, Joseph 39, 40, 49
Highway *128* 255–6, 259–60
Hollerith, Herman 84–5
Hughes, D.E. 50

IBM 78–9, 85
  computers 157, 159, 175, 176
ICL 225–6
incandescent lamps 68–9
industrial organizations 24, 25–6, 276–8
  importance of size 112–16, 274–5
  research spending 120, 235–6
information, defined 29
information technology, defined 28, 29
informational services 285–6
infrastructure 271, 273–4
innovations
  and governments 278–80
  and industry 276–8
  categories 22
  geography of 5–6, 260–1, 275–6
  role 4, 13, 16, 265–72
  societal 281–4
instruments, professional, scientific 210–12, Table 12.3
  employment, industrial 198–9, *12.2*
  patents Table 12.9, Table 12.10
integrated circuits 153, Table 9.1
International Business Machines *see* IBM
International Computers Limited (ICL) 225–6
International Telephone and Telegraph (ITT) 177, 180
internationalization 115

Japan 274, 275–6, 283
  employment, industrial 192–3, 195, 199–201
    electronics 197–8, 199, *11.5*
    office machinery 196
  production 163
    electronic goods 179, 180, 183–4, Table 10.4–.6
    office machinery 167
  research and development Table 12.11–.13, Table 12.15
  trade, international 204–5, 211–12
    electronic goods Table 10.5, 206–11
    office machinery 205

Kondratieff, Nikolai 3–4, 12–13
Kondratieff waves 15, Table 2.1, Table 2.2, 267–72
  second 37–54
  third 57–148
  fourth 151–261
  fifth 284–8

lighting, electric 60–1, 67–9
London 46, 145–8, 246–51
  western corridor 244–53
long wave theory 3–4, 12–16
  criticism 18–25
  New IT and 30–34, 268–72
  revival theory 16–18

magnetrons, 98–9
Marconi, Guglielmo Marchese 89–90, 91
Menlo Park 138–9
Mensch, Gerhard 17–18, 269, 270, 271
  theory criticism 18, 25
microcomputer industry 157–9, 258–9
microprocessors 153–4, 157
military expenditure 174, 182, 236–41, 279–80
minicomputers 157
Mix and Genest 133, 135, 136
Morse, Samuel 40–1

motors, electric 58, 70–1

National Telephone Company 51–2, 76
nationalization 42–4, 53, 136
New Information Technology
  defined 30, Table 11.1
  long wave theory 28–34, 268–72
New IT *see* New Information Technology
New York-Boston corridor 137–45, 254
NIC (Newly Industrializing Countries) 184, 213
  employment, industrial 191–5, 196, 197–8, 199
  trade, international 204–5, 207, 209, 210, 211

office machinery 76–86, 163–4, 167–73, 222–3
  employment, industrial 196, *12.2*
  production 167, 170–1, 222
  trade, international 78, 167–9, 205–6, Table 12.3
optical fibre transmission 161–2
Orange County, California 259–60

Parsons, Charles 66
patents 232–3
Perez, Carlota 20
phonograph 79–80
planar process 152–3, 257
Post Office 51, 52, 75–6, 95
power supply, electric 63–5
  London 114, 145
  see also d.c. generators
press-wire service 45
production 113, 287
  components, electronic 184–5, 228–30
    radio valves 108–9, 143, Table 9.1, 229
  electrical industry 76, 102, 106
  electronic capital goods 178–9
  employment and 100, 199–201, 218–21
  motors and early 70–1
  office machinery 167, 170–1, 222
  planar process 152–3, 257

radar 98–9, 174–5
radio 89–94
  equipment 107–10, 128, 177–8, 228
  detection and ranging *see* radar
RCA 93, 97–8, 142, 162–3
Remington Rand Corporation 85, 156, 157
research and development 93–4, 118–21, 234–40, 281, 282

Schumpeter, Joseph 4, 12, 13–16
  theory criticism 16
semiconductors 154, 181–3
  production 152–3, 185, 230–1
Siemens, Ernst Werner von 41–2, 117, 126, 134
Siemens & Halske 69, 103, 113, 127–9, 129–31
  employment Table 8.2
  research and development 134–5
Silicon Valley 256–60
software industry 159
sound film technique 129
sound recording 79–82, 206–9
South-East England 244–52
steam turbines 66–7
submarine telegraphy 44, 46
Swan, Sir Joseph Wilson 68–9, 111
switching equipment 160

tabulating-accounting machines 84–6
technology
  defined 29
  systems 266, 268–72
telecommunications 160–2, 179–81, 197–8, 206–9, *see also* telephone
Telefunken Company 92–3, 133
telegraph, electric 39–45, 128
  compared to telephone 44, 47–8
Telegraph Act (1869) 43, 51
telephone 48–50
  compared to telegraph 44, 47–8
  in Europe 50–2, 75–6
  in USA 49–50, 75
  long distance 75
television 96–8, 162–3, 208–9, 223
Tesla, Nikola 64–5
thermionic valves
  production 108–9, 143, Table 9.1, 229
  see also diode valves; triode valves
trade, international 104
  electronic goods 179, Table 10.5, 203–12
  office machinery 78, 167–9, 205–6, Table 12.3
  product groups defined Table 11.5
  see also under names of individual countries
tramways 69–70
transistors 151–3, 154, Table 9.1
triode valves 75, 88, 91
typewriters 76–9, Table 10.3

United Kingdom 279
  communications systems 40, 42–4, 51–2, 75–6, 95–6
    decline 100–22, 215–41
  employment, industrial *11.2*, 193, 195
    and output 199–202, 216–20
    electrical 76, 100, 104, 106

electronics 197, 199, *11.5*
office machinery 196
geography of innovation 5–6, 66–7, 275
in the future 287–8
patents 232–3
production 70–1, 111
electrical goods 102, Table 7.2, 106
electronic goods 177, 179, Table 10.4–.6
office machinery 167, 170
research and development 118–21, 234–41
trade, international 204–5, 211–12, 221–2
electrical goods 47, Table 7.4, 104
electronic goods 174, *11.11* – *11.14*
office machinery 167–9, 205
tramways 69, 70
see also London; South East England
United States of America 279–80
communications systems 40, 45, 49–50, 95
employment, industrial 192–3, 195, 253, 286
and output 199–202
electronics 144, 197, 198, 199
office machinery 196
geography of innovation 5–6, 275
in the future 283–4, 286–7
patents 232–3
production
electrical goods 102
electronic goods 109, 177, 178, 179–80, 181–5
office machinery 167
research and development 93–4, 119–20, Table 12.11, Table 12.12, Table 12.15
trade, international 204–5, 211–12
electronic goods Table 7.4, Table 7.8, 180, 206, *11.12*, 210
office machinery 205
tramways 69–70
see also Boston-New York corridor; Chicago; Silicon Valley
video cassette recorders 163
Volta pile 39

Western Union Telegraph Company 41, 49, 138
Westinghouse Electric Corporation Ltd 61, 62, 64, 65, 66, 94
Wiliams, Charles 49, 137, 140

xerography 164, 171–2